Current Topics in Microbiology and Immunology

Volume 324

Series Editors

Richard W. Compans
Emory University School of Medicine, Department of Microbiology and
Immunology, 3001 Rollins Research Center, Atlanta, GA 30322, USA

Max D. Cooper
Howard Hughes Medical Institute, 378 Wallace Tumor Institute, 1824 Sixth
Avenue South, Birmingham, AL 35294-3300, USA

Tasuku Honjo
Department of Medical Chemistry, Kyoto University, Faculty of Medicine,
Yoshida, Sakyo-ku, Kyoto 606-8501, Japan

Hilary Koprowski
Thomas Jefferson University, Department of Cancer Biology, Biotechnology
Foundation Laboratories, 1020 Locust Street, Suite M85 JAH, Philadelphia,
PA 19107-6799, USA

Fritz Melchers
Biozentrum, Department of Cell Biology, University of Basel, Klingelbergstr.
50–70, 4056 Basel Switzerland

Michael B.A. Oldstone
Department of Neuropharmacology, Division of Virology, The Scripps Research
Institute, 10550 N. Torrey Pines, La Jolla, CA 92037, USA

Sjur Olsnes
Department of Biochemistry, Institute for Cancer Research, The Norwegian
Radium Hospital, Montebello 0310 Oslo, Norway

Peter K. Vogt
The Scripps Research Institute, Dept. of Molecular & Exp. Medicine, Division of
Oncovirology, 10550 N. Torrey Pines. BCC-239, La Jolla, CA 92037, USA

Tatsuji Nomura • Takeshi Watanabe
Sonoko Habu
Editors

Humanized Mice

Tatsuji Nomura, M.D., Ph.D.
Central Institute for
  Experimental Animals,
Kawasaki,
Japan
tnomura@ciea.or.jp

Takeshi Watanabe, M.D., Ph.D.
RIKEN, Yokohama,
Japan
wtakeshi@rcai.riken.jp

Sonoko Habu, M.D., Ph.D.
Tokai University School of
  Medicine, Kanagawa,
Japan
sonoko@is.icc.u-tokai.ac.jp

ISBN 978-3-540-75646-0   e-ISBN 978-3-540-75647-7
DOI 10.1007/978-3-540-75647-7

Current Topics in Microbiology and Immunology ISSN 007-0217x

Library of Congress Catalog Number: 72-152360

© 2008 Springer-Verlag Berlin Heidelberg

This work is subject to copyright. All rights reserved, whether the whole or part of the material is concerned, specifically the rights of translation, reprinting, reuse of illustrations, recitation, broadcasting, reproduction on microfilm or in any other way, and storage in data banks. Duplication of this publication or parts thereof is permitted only under the provisions of the German Copyright Law of September, 9, 1965, in its current version, and permission for use must always be obtained from Springer-Verlag. Violations are liable for prosecution under the German Copyright Law.

The use of general descriptive names, registered names, trademarks, etc. in this publication does not imply, even in the absence of a specific statement, that such names are exempt from the relevant protective laws and regulations and therefore free for general use.

Product liability: The publisher cannot guarantee the accuracy of any information about dosage and application contained in this book. In every individual case the user must check such information by consulting the relevant literature.

Cover Design: WMXDesign GmbH, Heidelberg, Germany

Printed on acid-free paper

9 8 7 6 5 4 3 2 1

springer.com

Preface

The term humanized mouse in this text refers to a mouse in which human tissues and cells have been transplanted and show the same biological function as they do in the human body. That is, the physiological properties and functions of transplanted human tissues and cells can be analyzed in the mouse instead of using a living human body. It should therefore be possible to study the pathophysiology and treatment of human diseases in mice with good reproducibility. Thus, the humanized mouse can be used as a potent tool in both basic and clinical research in the future.

The development of appropriate immunodeficient mice has been indispensable in the creation of the humanized mouse, which has been achieved through many years of efforts by several laboratories. The first stage on the road to the humanized mouse was the report on nude mice by Isaacson and Cattanach in 1962. Thereafter, nude mice were studied in detail by Falanagan and, in 1968, Pantelouris found that these mice have no thymus gland, which suggested that the mice lack transplantation immunity against xenografts such as human hematopoietic stem cells. At the Nude Mouse Workshops (organized by Regard, Povlsen, Nomura and colleagues) that were held nine times between 1972 and 1997, the possibility of creating a humanized mouse using nude mice was extensively examined. The results, however, showed that certain human cancers can be engrafted in nude mice, but unfortunately engraftment of normal human tissue was almost impossible. Nevertheless, nude mice have made a great contribution to the evaluation of the effects of drugs against human cancer. In the course of this research, a system for production of a large number of germfree animals with the same high biological quality and know-how for strict microbiological and genetic quality control of the animals were established. It is now evident that highly reproducible research and experimental results could not be obtained without these achievements. This point should be taken into account in the future development of immunodeficient mice more suitable as humanized mice. Since human tissues other than certain cancer cells, especially hematopoietic cells, cannot be engrafted in nude mice, development of an even higher level of immunodeficient mice was needed. Great hopes were then placed on the SCID mice that have no T cells or B cells developed by Bosma in 1983, but the engraftment rate of normal tissue was again not as high as expected. Thereafter, NOD-scid mice were developed as an improved version of SCID mice (Shultz et al. and Ito et al.). Using NOD-scid mice, it became possible to achieve in vivo differentiation of human blood cells from a human hematopoietic stem cell, and these mice have been used in various applications. However, human cells can be engrafted only to a certain extent and adequate differentiation is not induced even in NOD-scid mice. Further development of new immunodeficient mice was still required. Recently, during 2000–2002, NOD-scid IL2R null (Ito et al., Schultz et al.) and Rag2nullIL2R null mice (Ito et al.) were developed. These mice have been proven to show much better engraftment and differentiation states than any other previously established mice and they have attracted the attention of researchers as a basis for creation of the humanized mouse. During the last few years, remarkable results have been reported concerning the establishment and the application of humanized mice by applying newly developed immunodeficient mice such as NOD-scid IL2R null and Rag2nullIL2R null mice. Based on recent progress, the first International Workshop on Humanized Mice was held in Tokyo in October 2006. This volume of Current Topics in Microbiology and Immunology is based mainly on the presentations at the workshop, but all chapters were especially written for the book. A wide range of topics is discussed in this volume including the characteristics of newly developed immunodeficient mice, their rearing conditions and points to consider, as well as the creation of humanized mice using various human cells and tissues, especially human hematopoietic stem cells, their characteristics, applications, and future prospects. We hope that the present volume will contribute to the expansion of research using the humanized mouse and the development of new humanized mice in the future.

Tatsuji Nomura
Takeshi Watanabe
Sonoko Habu

Contents

Basic Concept of Development and Practical Application of Animal Models for Human Disesases 1
T. Nomura, N. Tamaoki, A. Takakura, and H. Suemizu

Humanized SCID Mouse Models for Biomedical Research 25
T. Pearson, D. L. Greiner, and L. D. Shultz

NOD/Shi-*scid* IL2rγ^{null} (NOG) Mice More Appropriate for Humanized Mouse Models 53
M. Ito, K. Kobayashi, and T. Nakahata

Humanizing Bone Marrow in Immune-Deficient Mice 77
K. Ando, Y. Muguruma, and T. Yahata

The Differentiative and Regenerative Properties of Human Hematopoietic Stem/Progenitor cells in NOD-SCID/IL2rγ^{null} Mice.. 87
F. Ishikawa, Y. Saito, S. Yoshida, M. Harada, and L. D. Shultz

Antigen-Specific Antibody Production of Human B Cells in NOG Mice Reconstituted with the Human Immune System 95
R. Ito, M. Shiina, Y. Saito, Y. Tokuda, Y. Kametani, and S. Habu

Humanized Immune System (HIS) Mice as a Tool to Study Human NK Cell Development 109
N. D. Huntington and J. P. Di Santo

Human T Cell Development and HIV Infection in Human Hemato-Lymphoid System Mice 125
S. Baenziger, P. Ziegler, L. Mazzucchelli, L. Bronz, R. F. Speck, and M. G. Manz

Humanized Mice for Human Retrovirus Infection 133
Y. Koyanagi, Y. Tanaka, M. Ito, and N. Yamamoto

**Functional and Phenotypic Characterization of the
Humanized BLT Mouse Model** 149
A. K. Wege, M. W. Melkus, P. W. Denton, J. D. Estes, and J. V. Garcia

Novel Metastasis Models of Human Cancer in NOG Mice 167
M. Nakamura and H. Suemizu

In Vivo Imaging in Humanized Mice 179
H. Masuda, H. J. Okano, T. Maruyama, Y. Yoshimura,
H. Okano, and Y. Matsuzaki

Index 197

Contributors

K. Ando
Division of Hematopoiesis, Research Center of Regenerative Medicine,
Department of Hematology, Tokai University School of Medicine, Bohseidai,
Isehara, Kanagawa 259-1193, Japan, andok@keyaki.cc.u-tokai.ac.jp

S. Baenziger
Division of Infectious Diseases and Hospital Epidemiology, University Hospital
Zurich, Raemistrasse 100, 8091 Zurich, Switzerland

L. Bronz
Department of Obstetrics and Gynecology, Ospedale San Giovanni, 6500
Bellinzona, Switzerland

P. W. Denton
Department of Internal Medicine, Division of Infectious Diseases Y9.206,
University of Texas Southwestern Medical Center at Dallas, 5323 Harry Hines
Blvd., Dallas, TX 75390-9113, USA

J. P. Di Santo
Cytokine and Lymphoid Development Unit, Immunology Department, Institut
Pasteur, 25 rue du Docteur Roux, Paris 75724, France, disanto@pasteur.fr

J. D. Estes
Department of Internal Medicine, Division of Infectious Diseases Y9.206,
University of Texas Southwestern Medical Center at Dallas, 5323 Harry Hines
Blvd., Dallas, TX 75390-9113, USA
Department of Microbiology, Medical School, University of Minnesota,
MMC 196, 420 Delaware Street S.E., Minneapolis, MN 55455, USA

J. V. Garcia
Department of Internal Medicine, Division of Infectious Diseases Y9.206,
University of Texas Southwestern Medical Center at Dallas, 5323 Harry Hines
Blvd., Dallas, TX 75390-9113, USA, victor.garcia@utsouthwestern.edu

D. L. Greiner
Diabetes Division, University of Massachusetts Medical School,
373 Plantation Street, Suite 218, Worcester, MA 01605, USA

S. Habu
Department of Immunology, Tokai University School of Medicine, Bohseidai, Isehara, Kanagawa, Japan, sonoko@is.icc.u-tokai.ac.jp

M. Harada
Department of Biosystemic Medicine, Kyushu University Graduate School of Medical Sciences, Fukuoka, Japan

N. D. Huntington
Cytokine and Lymphoid Development Unit, Immunology Department, Institut Pasteur, 25 rue du Docteur Roux, Paris 75724, France

F. Ishikawa
Research Unit for Human Disease Model, RIKEN Research Center for Allergy and Immunology, 1-7-22 Suehiro-cho Tsurumi-ku, Yokohama, Kanagawa 230-0045, Japan, f_ishika@rcai.riken.jp

M. Ito
Laboratory of Immunology, Central Institute for Experimental Animals, 1430 Nogawa, Miyamae, Kawasaki 216-0001, Japan, mito@ciea.or.jp

R. Ito
Department of Immunology, Tokai University School of Medicine, Bohseidai, Isehara, Kanagawa, Japan

Y. Kametani
Department of Immunology, Tokai University School of Medicine, Bohseidai, Isehara, Kanagawa, Japan

K. Kobayashi
Mouse Mutation Resource Exploration Team, Functional Genomics Research Group, RIKEN GSC, 3-1-1 Koyadai, Tsukuba, Ibaraki 305-0074, Japan

Y. Koyanagi
Laboratory of Viral Pathogenesis, Institute for Virus Research, Kyoto University, 53 Shougoinkawara cho, Sakyou-ku, Kyoto 606-8507, Japan, ykoyanag@virus.kyoto-u.ac.jp

M. G. Manz
Institute for Research in Biomedicine (IRB), Via Vincenzo Vela 6, 6500 Bellinzona, Switzerland, manz@irb.unisi.ch

T. Maruyama
Department of Obstetrics and Gynecology, Keio University School of Medicine, 35 Shinanomachi, Shinjuku-ku, Tokyo 160-8582, Japan

H. Masuda
Department of Obstetrics and Gynecology, Keio University School of Medicine, 35 Shinanomachi, Shinjuku-ku, Tokyo 160-8582, Japan

Y. Matsuzaki
Department of Physiology, Keio University School of Medicine,
Tokyo 160-8582, Japan

L. Mazzucchelli
Institute for Pathology, 6600 Locarno, Switzerland

M. W. Melkus
Department of Internal Medicine, Division of Infectious Diseases Y9.206,
University of Texas Southwestern Medical Center at Dallas,
5323 Harry Hines Blvd., Dallas, TX 75390-9113, USA

Y. Muguruma
Division of Hematopoiesis, Research Center of Regenerative Medicine,
Department of Hematology, Tokai University School of Medicine,
Bohseidai, Isehara, Kanagawa 259-1193, Japan

T. Nakahata
Department of Pediatrics, Graduate School of Medicine, Kyoto University,
54 Kawaharacho, Shogoin, Sakyo-ku, Kyoto 606-8507, Japan

M. Nakamura
Central Institute for Experimental Animals, 1430 Nogawa, Miyamae,
Kawasaki 216-0001, Japan, nakamura.masato@hachioji-hosp.tokai.ac.jp

T. Nomura
Central Institute for Experimental Animals, Nogawa 1430, Miyamae,
Kawasaki 216-0001, Japan

H. Okano
Department of Physiology, Keio University School of Medicine,
Tokyo 160-8582, Japan

H. J. Okano
Department of Obstetrics and Gynecology, Keio University School of Medicine,
35 Shinanomachi, Shinjuku-ku, Tokyo 160-8582, Japan

T. Pearson
Diabetes Division, University of Massachusetts Medical School,
373 Plantation Street, Suite 218, Worcester, MA 01605, USA

Y. Saito
Department of Breast and Endocrine Surgery, Tokai University
School of Medicine, Bohseidai, Isehara, Kanagawa, Japan

Y. Saito
Research Unit for Human Disease Model, RIKEN Research Center for Allergy
and Immunology, 1-7-22 Suehiro-cho Tsurumi-ku, Yokohama,
Kanagawa 230-0045, Japan

M. Shiina
Department of Immunology, Tokai University School of Medicine,
Bohseidai, Isehara, Kanagawa, Japan

L. D. Shultz
The Jackson Laboratory, 600 Main Street, Bar Harbor, ME 04609,
USA, lenny.shultz@jax.org

R. F. Speck
Division of Infectious Diseases and Hospital Epidemiology,
University Hospital Zurich, Raemistrasse 100, 8091 Zurich, Switzerland

H. Suemizu
Central Institute for Experimental Animals, Nogawa 1430, Miyamae,
Kawasaki 216-0001, Japan

A. Takakura
Central Institute for Experimental Animals, Nogawa 1430, Miyamae,
Kawasaki 216-0001, Japan

N. Tamaoki
Central Institute for Experimental Animals, Nogawa 1430, Miyamae,
Kawasaki 216-0001, Japan, tnomura@ciea.or.jp

Y. Tanaka
Department of Immunology, Graduate School of Medicine,
University of the Ryukyus, Okinawa 903-0215, Japan

Y. Tokuda
Department of Breast and Endocrine Surgery, Tokai University
School of Medicine, Bohseidai, Isehara, Kanagawa, Japan

A. K. Wege
Department of Internal Medicine, Division of Infectious Diseases Y9.206,
University of Texas Southwestern Medical Center at Dallas,
5323 Harry Hines Blvd., Dallas, TX 75390-9113, USA

T. Yahata
Division of Hematopoiesis, Research Center of Regenerative Medicine,
Department of Hematology, Tokai University School of Medicine,
Bohseidai, Isehara, Kanagawa 259-1193, Japan

N. Yamamoto
AIDS Research Center, National Institute of Infectious Diseases,
Tokyo 162-8640, Japan

S. Yoshida
Research Unit for Human Disease Model, RIKEN Research Center for Allergy
and Immunology, 1-7-22 Suehiro-cho Tsurumi-ku, Yokohama,
Kanagawa 230-0045, Japan

Y. Yoshimura
Department of Obstetrics and Gynecology, Keio University School of Medicine, 35 Shinanomachi, Shinjuku-ku, Tokyo 160-8582, Japan

P. Ziegler
Institute for Research in Biomedicine (IRB), Via Vincenzo Vela 6, 6500 Bellinzona, Switzerland

Basic Concept of Development and Practical Application of Animal Models for Human Diseases

T. Nomura(✉), N. Tamaoki, A. Takakura, and H. Suemizu

1	Introduction	2
2	Basic Concept of Development and Practical Application of Animal Models for Human Diseases and *In Vivo* Animal Experimentation Systems	3
	2.1 Establishment as Standardized Laboratory Animals (Stage 1)	3
	2.2 Establishment as a Disease Model (Stage 2)	7
	2.3 *In Vivo* Experimentation Systems (Stage 3)	7
3	The Humanized Mouse	8
4	NOG Mouse That Serves as the Basis for Humanized Mice and Their Controls	8
	4.1 Basic Concept of Establishing the NOG Mouse as a Standardized Laboratory Animal	9
	4.2 Significance of Genetic Quality Control of NOG Mice	9
	4.3 Significance of Environmental Control of NOG Mice	15
5	Future Aspects of Humanized Mice	21
6	Conclusion	21
References		22

Abstract The "humanized mouse" is a mouse harboring functioning human tissues used as *in vivo* human models for both physiological and pathological conditions. The NOD/Shi-*scid IL2rγ^{null}* (NOG) mouse, an excellent immunodeficient mouse used as the basis for the humanized mouse, requires strict genetic and environmental control for production and use in experiments. Genetic control using marker-assisted selection is described. In addition, NOG mice are easily affected by microbiological and proximate environmental factors, which can cause severe damage to the mice in some cases. Therefore, rigorous

T. Nomura
Central Institute for Experimental Animals, Nogawa 1430, Miyamae,
Kawasaki 216-0001, Japan
tnomura@ciea.or.jp

microbiological and environmental controls are necessary to ensure reproducibility of experimental results. At the end of this chapter, future aspects of the application of "humanized mice" based on novel super-immunodeficient mice such as NOG mice and $Rag2^{null}\ IL2r\gamma^{null}$ mice in biomedical research and testing are briefly reviewed.

Abbreviations EGFP: enhanced green fluorescent protein; ES: embryonic stem cell; HIV-1: human immunodeficiency virus type 1; IgA: immunoglobulin A; IL: Interleukin; MHV: mouse hepatitis virus; PVR: poliovirus receptor; Rag2: recombination activating gene 2; Tg: transgenic

1 Introduction

The purpose of this review is to give an overall and clear picture of the humanized mouse and the animal models on which it is based.

The latest research using laboratory animals tends to be *in vivo* studies on function of molecules since the establishment of methods to create transgenic and knockout mice with molecular biology techniques. Molecular biologists, however, consider only the molecular function and often overlook the genetic background and environmental factors such as microbiological factors that affect molecular function. This approach makes it difficult to obtain reproducible experimental results and, often, to evaluate the results obtained in biomedical research.

In this review, basic concepts for how to create animal models for biomedical research on human disease and how to control environmental factors that affect *in vivo* experimental systems that use these animals are presented.

Immunodeficient mice that form the basis for the humanized mouse have been a major research theme in our institute, the Central Institute for Experimental Animals (CIEA), since nude mice were introduced in 1974. The humanized mouse is very useful in clarification of the mechanisms and treatment of human diseases directly with laboratory animals.

After various attempts to establish immunodeficient mice, we were recently successful in establishing the NOG immunodeficient mouse, which is ideal as a humanized mouse. The first step was to achieve genetic control by specifying genetic quality standards. We found, however, that NOG mice often contracted serious diseases during experiments because of environmental factors including microbiological factors, and stable experiments could not be performed. The pathology of these diseases and how they should be controlled are discussed below. This is very important information for the use of NOG mice as humanized mice.

2 Basic Concept of Development and Practical Application of Animal Models for Human Diseases and *In Vivo* Animal Experimentation Systems

The basic concept of the establishment of laboratory animals as standardized human disease models is explained in Fig. 1. It is divided into three basic stages. Immunodeficient NOD/Shi-*scid IL2Rγ^{null}* (we called this mouse "NOG") mice, which form the basis of the humanized mouse, were also developed based on this concept.

2.1 *Establishment as Standardized Laboratory Animals (Stage 1)*

In Stage 1, establishment as standardized laboratory animals, three steps must be completed.

2.1.1 Establishment of Quality Standards (Step 1)

The first step is the establishment of quality standards or specifications for the animal models. Reproducibility cannot be ensured if there are no quality standards for the animals. The establishment of standards is the starting point for establishment as a standardized laboratory animal. Quality standards specify genetic quality that defines the phenotype and microbiological quality that defines the dramatype. The definitions of the phenotype and dramatype are as follows.

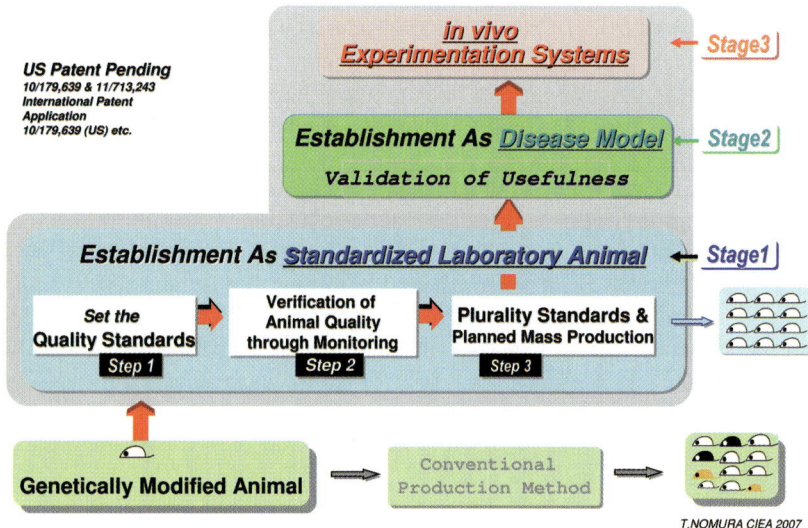

Fig. 1 Basic concept of human disease model and integrated *in vivo* experimental systems

The term "phenotype" refers to the "outward physical manifestation" of a living organism. These are the physical parts, cell structures, metabolism, energy utilization, tissues, organs, reflexes, and behaviors: anything that is part of the observable structure and function of behavior of the organism.

The term "dramatype," which was originally proposed by Russell and Burch in 1959 [32], refers to the pattern of performance in a single physiological response of a laboratory animal. Variation in such responses is the joint product of two factors, the phenotype itself and the proximate environment in which the animals are tested such as temperature, humidity, diet, investigators, and animal care personnel. For a uniform dramatype, the environmental conditions in which the animal experiments are performed must be strictly controlled. The relation between quality standards, reproducibility, and the dramatype is outlined below.

Quality Standards and Assurance of Reproducibility in Animal Experiments

It has been noted by many researchers that even when genetically identical mice from the same source, that is, those with an identical genotype, are used in different laboratories for medical research such as cancer research to examine the occurrence of spontaneous tumors, incidence of induced tumors, and cancer chemotherapy screening, conflicting results are often obtained. It is speculated that that environmental factors are the main cause of these discrepancies.

A wide variety of evidence is available showing that experimental results are in fact affected by environmental factors such as temperature, humidity, barometric pressure, lighting, season, diet, noise, odors, crowding, confinement, group composition, and contact with other animals and humans. Of these factors, temperature has been the most widely investigated. For example, it is reported that drug toxicity shows wide differences with changes in environmental temperature [29, 30, 36, 37]. In microbiological and immunological experiments, the experimental results are strongly affected by environmental temperature. Therefore, it is evident that environmental factors should be strictly controlled to obtain reliable and reproducible results in animal experiments.

To this end, it is essential to understand the basic relationship between animal experiments and environmental factors. These basic factors should be considered in terms of the concept of the dramatype. The original definition of the dramatype by Russell and Burch is rarely considered in current biomedical research using laboratory animals. Therefore, we propose a new definition of the dramatype from the standpoint of animal experimentation in biomedical research. It was once generally considered that the results of animal experiments depended only on the response of the phenotype, but this is not the case. In actual animal experiments in which environmental factors are important, it becomes necessary to consider a new concept, the dramatype, which is determined by the combined action of two factors: the phenotype and the proximate environment (Fig. 2). Any experimental result using animals should be recognized as the response of the dramatype to the experimental procedure.

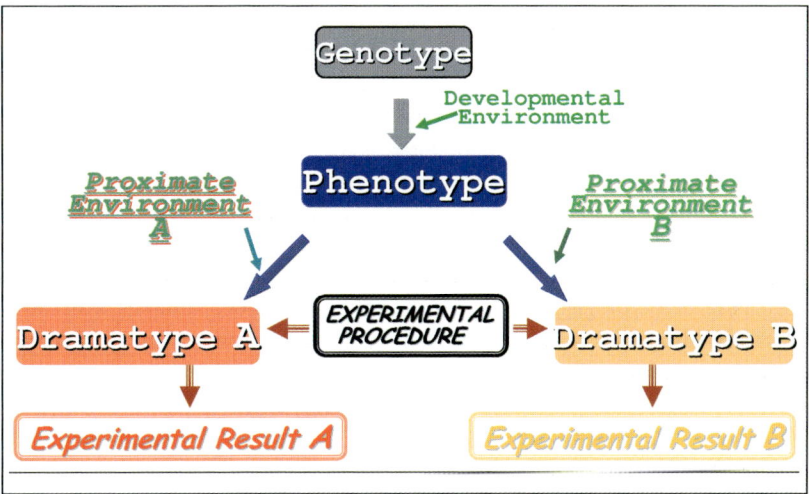

Fig. 2 Proximate environment can cause differences in dramatype and experimental results

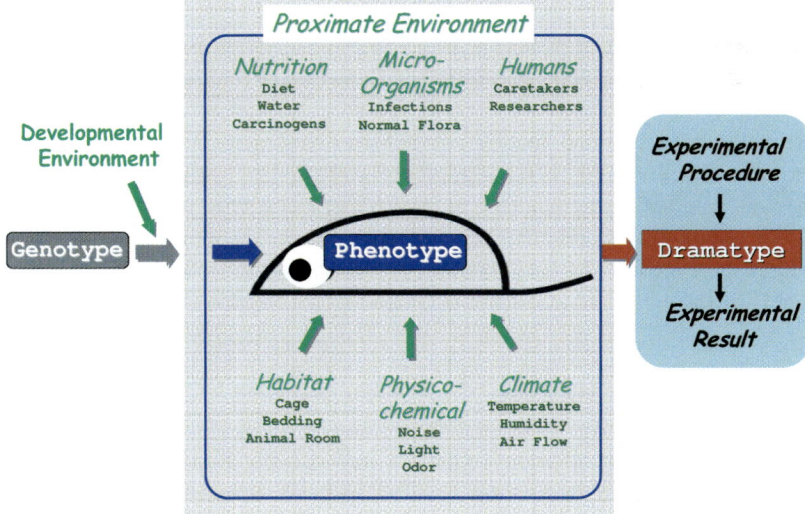

Fig. 3 Key factors in determining dramatype

Figure 3 represents the reciprocal relationship among "genotype," "phenotype," and "dramatype." The only factor that affects phenotype is genetic quality; on the other hand, there are many environmental factors that affect dramatype. Genetic quality control of the NOG mouse is explained in Sect. 4.2, and environmental and physiological quality controls are explained in Sect. 4.3.

2.1.2 Establishment of Monitoring System for Verification of Compliance with Quality Standards (Step 2)

The second step is the establishment of a monitoring system that verifies that the quality standards are properly maintained in the animals produced. Both genetic and microbiological testing should be conducted on a scheduled basis.

The term "scheduled monitoring" refers to an examination that is performed regularly by standard methods to ensure that the genetic and microbiological quality of the defined animal is stable over time. This can be achieved by comparing the genetic and microbiological profiles of the defined (founder) animals with those of the corresponding inbred strains. This scheduled monitoring ensures not only animal health but also the specified quality [26, 27, 39].

Biological functions of animal models that serve as *in vivo* measurement tools change day by day, contrary to the tools used in physicochemical measurements. Monitoring serves as a sort of calibration to ensure that no functional changes have occurred in the animals as measuring instruments. The first two steps are important in determining whether or not the animals produced have reproducibility of biological functions.

2.1.3 Establishment of Mouse Mass Production System Based on Plurality Standards Required for Laboratory Animals (Step 3)

Since we consider animal models for human diseases to be *in vivo* measuring instruments, they should be mass-produced with defined plurality standards and their biological functions should be monitored periodically to verify that they comply with the quality standards. This ensures reproducibility in experiments and validates the reliability of the procedures studied. To this end, the following two systems are used.

Planned Production of Mice Based on Plurality Standards

Plurality standards for laboratory animals are critically important for bioassays since several identical animals should be used in an animal experiment with statistical analysis to obtain reliable results. The plurality standards are also required whenever and wherever animal experiments are performed.

Cryopreservation of Mouse Fertilized Embryos for Breeding Stocks

Cryopreservation preserves the fertilized embryos of the original breeding (founder) stock as a reference for genetic standards. It is also one aspect of risk management for dealing with unexpected accidents such as genetic or microbiological contamination.

The third step is based on the plurality standards, standards that cover many animals, rather than on the standards for each individual animal. In safety and efficacy studies on medicinal products, which include various factors such as dose and route of administration, it is necessary to use many animals with the same standards. Therefore, a reliable mass production system of animals with the same standards is indispensable.

In the case of the genetically modified TgPVR21 mouse developed by CIEA and approved for use in neurovirulence testing of live polio vaccine, a total of 52,000 mice with the same standards have been supplied by CIEA in the past 10 years to 10 research institutes throughout the world. CIEA developed these mice in close collaboration with WHO [4, 5, 17].

2.2 Establishment as a Disease Model (Stage 2)

Stage 2 is the establishment of a validation system to verify the usefulness and limitations of new mass-produced mice with plurality standards for medical research. Research exploring the basic mechanism of diseases, treatment and prevention methods, and drug discovery are meaningful only when validated animals are used. Validation of the disease model is performed by researchers with a detailed knowledge of the actual disease to ensure that the model has identical biological functions for the targeted disease. Once the validation has been completed, the animals can be treated as a standardized disease model. Close collaboration should, therefore, be maintained with the biomedical community as the users of the animals.

2.3 In Vivo Experimentation Systems (Stage 3)

In Stage 3, an *in vivo* experimentation system is established to provide the bridge from bench to bedside, that is, from basic research to practical application. These systems include proximate environmental control and animal experimentation methods such as the intraspinal inoculation method used in the neurovirulence test for oral polio vaccine, which is explained in more detail below.

In the studies on the use of transgenic mice to replace monkeys in the polio vaccine neurovirulence test, various routes of administration including oral, pernasal, percutaneous, subcutaneous, intradermal, intravenous, and intraspinal routes were examined with different doses in four strains of transgenic mice. Surprisingly, it was found that the least sensitive standardized laboratory animal for any practical application. In the case of TgPVR mice, it was necessary to first complete the *in vivo* experimentation system, followed by establishing the TgPVR21 mouse model as a standardized laboratory animal.

These three stages described above are the essential processes in the use of laboratory animals as human disease models in studies on the etiology of diseases and the development of new drugs.

3   The Humanized Mouse

Based on this concept of development and practical application of animal models for human disease, we have attempted to develop various animal models to date. One of our activities is the development of immunodeficient mice in which human cells, tissues, and organs grow and function as in a human being. This is the "humanized mouse" in which human diseases and biofunctions are reproduced.

Recently, however, the term "humanized mouse" has been used inappropriately because of insufficient understanding of its actual meaning. We are afraid that this will cause even greater confusion in the future if it is not clarified. Therefore, we must clearly define the term "humanized mouse." The "humanized mouse" is a mouse in which normal or abnormal human tissue that retains human biofunction can be transplanted and the biofunctions can be observed by medical researchers from a clinical standpoint. It is still a mouse but it maintains the identical biofunctions of human tissues such as immunological functions. The NOG mouse, in which human umbilical cord blood $CD34^+$ cells are transplanted, is an example in that human cells are differentiated and a human immune system can be reconstructed as in the case of human immunodeficiency virus (HIV) infections [12]. Another example is functional human endometrium regenerated from singly dispersed human endometrial cells in the kidney capsule of NOG mice. In NOG mice with human endometrium, hormone-dependent changes, including proliferation, differentiation, and tissue breakdown and shedding (menstruation), can be reproduced [19]. The "humanized mouse" is also a mouse in which human genes can be introduced so that the mouse has an identical function as defined by the introduced gene to that of humans [e.g., symptoms of paralysis induced by poliovirus infection in TgPVR] [11, 14].

4   NOG Mouse That Serves as the Basis for Humanized Mice and Their Controls

This is the basic concept of human disease models and *in vivo* animal experimentation systems at CIEA. A patent on CIEA's basic concept of the development of animal models for human disease is now pending in the US Patent Office. In China, the patent was granted under the title "Method for Developing Animal Models" (Chinese Patent No. 3820050.3 granted on March 2, 2007).

The TgPVR21 mouse [5, 12, 17] and the TgrasH2 mouse[21, 25, 41], which are distributed around the world, are two human disease models established from genetically modified mice using the basic concept of CIEA. CIEA developed the NOG mouse based on this concept.

4.1 Basic Concept of Establishing the NOG Mouse as a Standardized Laboratory Animal

Scientists in the fields of mammalian genetics and molecular biology who use mice in their research are not usually interested in plurality quality standards, monitoring to verify quality, or mass production systems for laboratory animals. However, it is highly desirable that quality standards or specifications of animals to be used in medical research are established and scheduled monitoring is performed to verify that such quality is maintained in animal models. If reproducibility of the results of animal experiments cannot be ensured, then such experiments cannot be recognized as scientific methodology. In bioassays, sufficient numbers of mice with the same standards should be used in each study, and plurality standards are required to ensure that all the mice have the same quality standards.

4.2 Significance of Genetic Quality Control of NOG Mice

In this section, the genetic quality control of "humanized mice" that were created from the NOG strain is described. In the first half of this section, the significance of genetic quality control in laboratory animals is outlined, and in the latter half genetic quality control and efficient improvement of NOG mice are described.

Recently, research using genetically modified mice has become common in all fields of biomedical research. Transgenic mice are generated by the introduction of the gene of interest into pronuclear cells of the fertilized eggs and established as transgenic lines with the introduced gene stably incorporated into the host mouse genome. Gene knockout mice have the gene of interest replaced with a selective marker gene, and gene-targeted events with homologous recombination must be performed with embryonic stem cells (ES cells). Since genetic uniformity is an important consideration, both genetically modified mice are usually generated in an inbred strain background.

4.2.1 Genetic Background Affects Phenotype

A genetically modified mouse with an inbred strain background is usually generated for specific purposes. The main purpose for producing genetically modified mice is an *in vivo* functional analysis of the "gene involved in the research." However, some of these animals are established as laboratory animals that serve as models for human diseases. At present, genetically modified mice with various genetic backgrounds are being established. The important point is that the phenotypes of each inbred strain differ among strains. When platelet functions were compared in three commonly used inbred strains, the tail bleeding times of BALB/c mice were reported to be statistically longer than those in the 129Sv and C57BL/6 strains [42]. Another example, the C3H/HeN(C3H) mouse strain, shows higher incidences of spontaneous and chemically induced hepatocellular carcinoma than the C57BL/6 strain [3, 6]. In the C3H strain, it has been reported that hepatocellular carcinoma induced by diethylnitrosoamine (DEN) is larger in size and the number of hepatocellular carcinomas detected by light microscopy is greater than in the C57BL/6 strain [16]. If transgenic mice, which possess a carcinogenesis-related gene, are produced in the C3H strain, the genetic background of the C3H strain will cause interpretation of carcinogenetic rate data in the liver to become complex in carcinogenicity testing. When the genetic background affects the phenotypes of such introduced genes, genetically modified mice with a more appropriate genetic background should be generated. However, even if transgenic mice are generated again with the C57BL/6 strain, the integration sites of the extrinsic genes will differ and the position effect will also differ, making it difficult to obtain exactly the same transgenic line. For the above transgenic mice with carcinogenesis-related genes, one effective method is to develop a congenic strain with the C57BL/6 strain that has a relatively lower incidence of spontaneous carcinoma.

The NOD-*scid* mouse that is now widely used as an immunodeficient mouse originated from the C.B-17 strain (similar to the BALB/c strain except that it carries the Igh-1b allele from the C57BL/Ka strain), which was discovered by Bosma et al. of the Fox Chase Cancer Center in 1983 [1]. Koyanagi et al. compared HIV-1 replication with the human hematopoietic cell reconstituted C.B17-*scid* and NOD/Shi-*scid* (*scid* mutation was introduced into the NOD/Shi strain by backcrossing) mice and reported that replication occurred at higher levels in NOD/Shi-*scid* mice and viremia appeared only in NOD/Shi-*scid* mice [15]. Such a genetic background is an important factor that directly affects character expression of the phenotypes in both wild-type and genetically modified animals. Replacement of the genetic background was performed as an effective method of improving the breed in order to generate laboratory animals most suited to the research objective. Genetic background and gene functions have a potentiative action, and examples of phenotype control are shown below. Engraftment of the human T-cell lymphoma cell line LM-2-JCK was compared in three strains: C.B17-*scid*, NOD/Shi-*scid* with only a different genetic background, and NOG mice with the same genetic background as NOD/Shi-*scid* and with the *IL2Rγ^{null}* mutation (Fig. 4). C.B17-*scid* and NOD/Shi-*scid* mice showed differences in tumor engraftment because of differences in genetic background even though they both have the *scid* mutation, and it was evident that the genetic background of the NOD/Shi strain shows higher immunodeficiency. This was considered to be caused by multiple defects of congenital and acquired immunity such as NK cell function incompetence, lack of complements, and differentiation and functional defects of antigen-presenting cells in the NOD/Shi strain. The IL-2 receptor (IL2-R) common γ-chain, which is a common subunit used in many cytokine receptors such as IL-2, IL-4, IL-7, IL-9, IL-15, and IL-21, has been destroyed in the NOG mouse with the *scid* mutation on the NOD/Shi genetic background, and it resulted in loss of functions. Therefore, the NOG mouse has both the T-cell and B-cell defects characteristic of NOD/Shi-*scid* mice

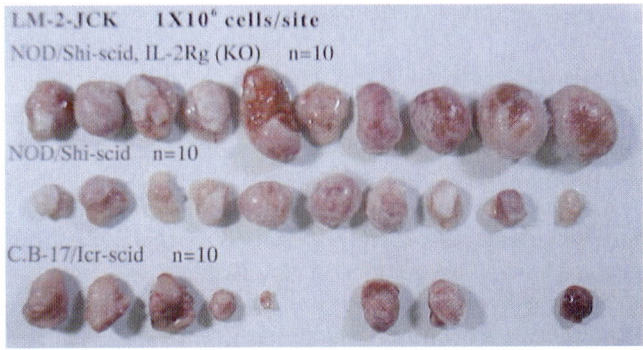

Fig. 4 Engraftment of human T-cell lymphoma cell line LM-2-JCK was compared in three immunodeficient mouse strains

and also shows NK-cell defects and multiple immune function disorders associated with lack of production of cytokines important in immune response [34]. The results of an experiment on transplantation of the human T-cell lymphoma cell line, LM-2-JCK, in NOG mice are shown in the top part of Fig. 4. Engraftment of tumor cells was observed in all mice in the same way as in NOD/Shi-*scid* mice, but the size of the tumor mass, that is, the growth, was better in NOG mice than in NOD/Shi-*scid* mice.

4.2.2 Necessity of Genetic Quality Standards (Phenotypic Control)

These genetically modified animals with confirmed characteristics as animal models are developed as laboratory animals, that is, mating methods are aimed at mass production, methods of testing the modified gene and the genetic background are established, and standards or specifications for evaluation of the test results are set as the genetic quality standards of the animal. NOG (NOD/Shi-*scid IL2Rγ^{null}*) mice are produced by mating a female NOD/Shi-*scid IL2Rγ^{null}* and a male NOD/Shi-*scid IL2Rγ^{null}* mouse. Since all of the mice produced are NOG mice, testing of the modified gene is not required.

When phenotype analysis is performed as a bioassay tool, scheduled genetic background testing or monitoring should be conducted because genetic identity is required. The Tg.AC transgenic mouse has multiple copies of the transgene *v-Ha-ras* at single loci on the chromosome. In this mouse, it has been reported that loss of the transgene in the palindromic sequence due to a repeat formation of the transgene occurred and the phenotype disappeared [38]. In scheduled monitoring to prevent such accidents, it is recommended to perform a dynamic phenotype analysis including biochemical, pathological, and immunological tests as well as static molecular biology tests.

4.2.3 Improvements of the NOG Mouse

In this section, improvements of the NOG mouse are described. The methods of introducing additional genetic changes in NOG mice that have already been completed include (1) the transgenic method of direct microinjection of the fertilized eggs of NOG mice and (2) the congenic method using existing genetically modified mice. The congenic method is the classical method that is easiest to perform. The targeted animal can be reliably obtained by backcrossing with the NOG mouse because phenotype expressions of the modified gene in the donor have already been identified. However, with this method, backcrossing must be performed for at least eight generations to completely replace the genetic background (99.6% is replaced theoretically), and this takes a long time. One of the fastest ways to introduce additional genetic modification is the marker-assisted selection protocol known as the "speed congenic" method. Since most of the knockout mice produced at present are generated by using ES cells with the 129 strain, the genetic background is often replaced by that of the C57BL/6 strain. Therefore, a microsatellite marker panel with differences between the 129 strain and the C57BL/6 strain has been developed [7]. In the marker-assisted selection protocol, the male mouse showing the most complete replacement with the targeted genetic background is selected as the "best male" and used in the next backcross [40]. Figure 5 shows the concept. We independently developed a panel of 64 microsatellite markers showing differences between the 129 strain and the C57BL/6 strain and also developed a panel of 64 microsatellite markers showing differences between the C57BL/6 strain and the NOG strain (manuscript in preparation). In improvement of the NOG mice, these 64 markers serve as genetic quality standards and the final backcross is completed when these markers become NOG markers (however, there are cases in which the introduced gene or vicinity of the altered gene do not become NOG). An example of an improved NOG mouse generated by the speed congenic method is presented here. The enhanced green fluorescent protein (EGFP) gene in C57BL/6-Tg (Act-EGFP) C14-Y01-FM1310sb (C57BL/6J TgEGFP) [23] was introduced into the NOG genetic background by the speed congenic method. The first generation hybrid (F_1) was produced by mating a male C57BL/6J TgEGFP mouse as donor with the recipient NOG female mouse (Fig. 6a). The F1 hybrid obtained received a uniform genome (except for the sex chromosome) from the donor and recipient, and the replacement rate, both the theoretical and measured values, was 50%. The F_1 hybrid is useful as a laboratory animal because all F_1 mice have the same genetic background after combining the characteristics of the two strains. Animals with the same genetic background (genetic quality) can be produced in any number at any time as required since these F_1 hybrids can be reproduced. The same genetic quality and reproducible production are not only the basic concept but also indispensable conditions for laboratory animals. In the next stage, randomly selected male F_1 Tg-EGFP mice are again mated with female NOG recipient mice to obtain the N_2 generation. Marker-assisted selection is performed for the N_2 generation male Tg mice. Using 64 markers, the male showing the closest NOG strain type is selected as the parent

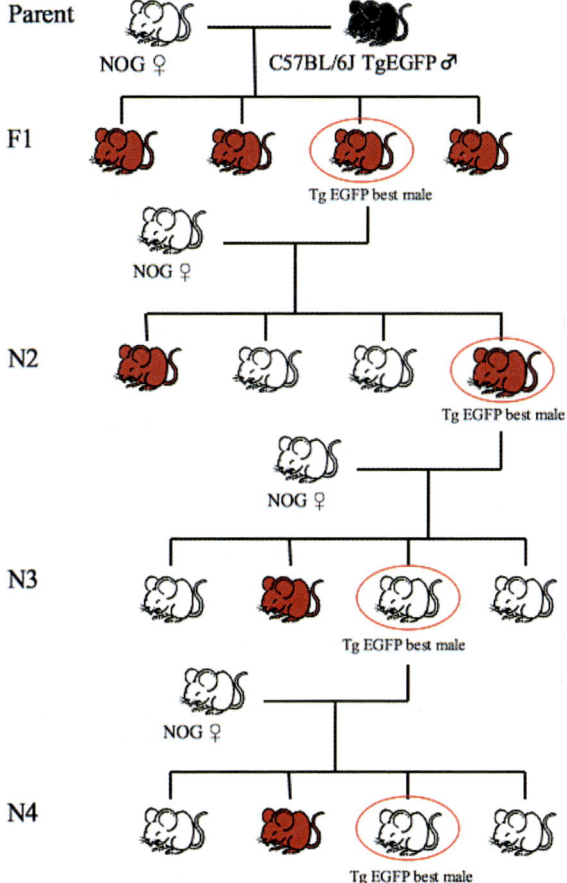

Fig. 5 Schematic representation of marker-assisted selection protocol known as "speed congenic." The "best male" enclosed in a *red circle* was used in the next backcross

for the next generation (Fig. 6b). The most important point to remember here is that all animals in the N_2 generation with genetic diversity due to homologous recombination ("crossing over") at the time of meiosis have different genetic backgrounds. Because the N_2 generation mice cannot be produced as animals with the same genetic backgrounds as in the case of inbred strains or F_1 hybrids, they should not be used as animal models in bioassays. With this method, selection of males with the closest NOG strain type was repeated (Fig. 6c). Finally, we were able to obtain an NOG mouse with the improved genetic background by backcrossing four times (Fig. 6d). This mouse cleared the minimum genetic quality standards consisting of 64 markers studied with the mouse genome for improved NOG mice. This mouse was used as the founder for the next N_5 backcross generation.

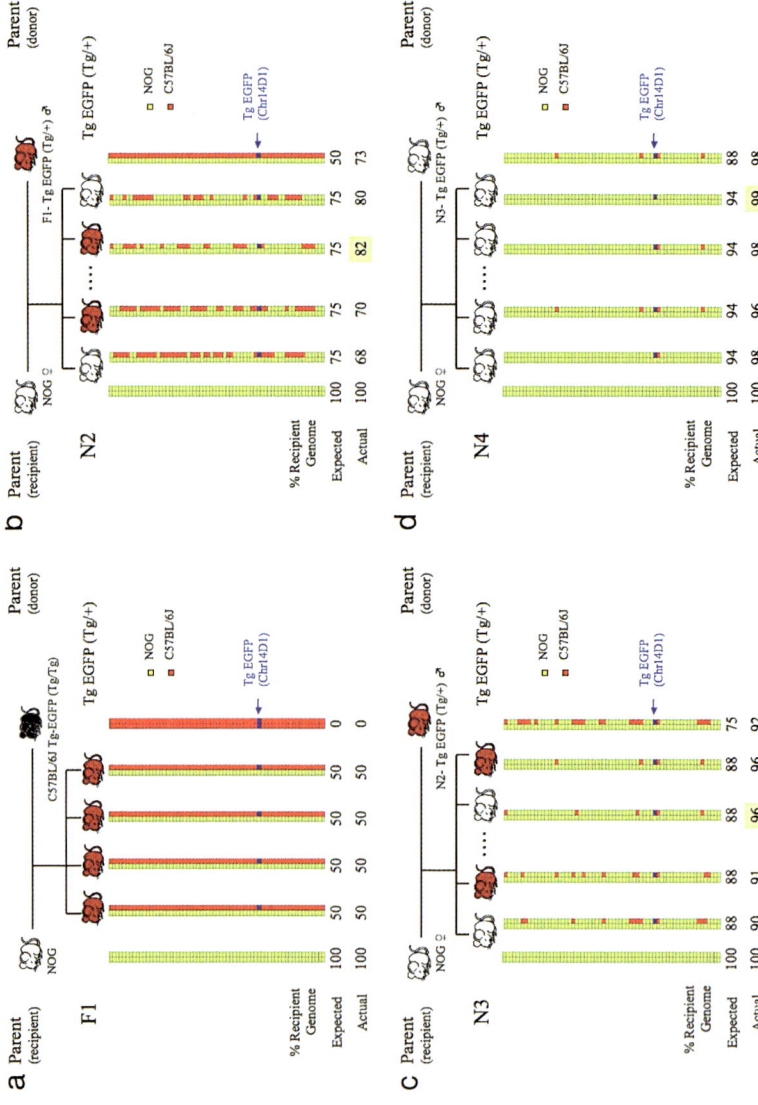

Fig. 6 Typical genetic profiles created by 64 microsatellite markers at each generation. The differences between the C57BL/6J and NOG strains are shown by colored boxes. *Yellow* and *red boxes* indicate the NOG and C57BL/6J genotype of the informative markers, respectively. Two rows correspond to the diploid state

The importance of genetic quality control and the genetic quality standards for "humanized mice" generated with the NOG mouse has been described. For animal models in bioassays, we must achieve reproducible production of animals with the same genetic background (genetic quality) and mass production with plurality standards. Therefore, the microsatellite markers that we set are the minimum genetic quality standards, and mice complying with these standards are improved NOG mice.

4.3 Significance of Environmental Control of NOG Mice

With NOG mice, which are known to have major differences in resistance to stress and infections from conventional immunodeficient mice, it is not sufficient simply to specify standards for the animals to ensure reproducibility of the results of animal experiments. It is also indispensable to control environmental factors, including microbiological factors, which have critical effects on the results of animal experiments. This has become clear from experience over the past 4 years and was not found with conventional immunodeficient animals. Therefore, control of environmental factors in sites where animal experiments are performed is especially important.

4.3.1 Immunodeficient Animals and Microbiological Factors

Since the appearance of the nude mouse, which was shown to be immunodeficient and able to be engrafted with human cells by Rygaard and Polvsen [33] in 1969, various immunodeficient animals such as the nude rat and *scid*, NOD-*scid*, and Rag2KO mice have been generated. These animals have made very important contributions to immunological and cancer research [1, 2]. However, a constant struggle is waged against microbial infections present in the rearing environment including infections with mouse hepatitis virus (MHV) and *Streptococcus aureus* in nude mice and *Pneumocystis carinii* (*P. carinii*) in *scid* mice, which, except for MHV, do not need to be controlled for animals with normal immune function [28, 31]. This problem can be solved by establishing a system to prevent infections including cleaning of infected animals with germ-free breeding techniques and development of new rearing facilities. It has now become possible to use immunodeficient animals of high microbiological quality safely [28]. In other words, the generation of immunodeficient animals has greatly improved the microbiological quality control system of laboratory animals.

The development of techniques for generation of genetically modified animals in recent years has undergone a major change in methods for development of human disease models established from conventional mutants, and the generation of animal models for various human diseases is easier than before. Many animals have been improved with genes of the immune system, and in the future researchers will generate a variety of severely immunodeficient animals resembling the

NOG mouse, which will contribute to advances in immunology, infectious diseases, and oncology.

Severely immunodeficient animals such as NOG mice produced by these techniques were considered impossible to produce in the past. Their resistance is weaker than that of existing immunodeficient animals, that is, they are severely immunodeficient animals. Therefore, their adaptability to conventional rearing and animal experimentation environments is unknown. In the 4 years since the development of the NOG mouse we have experienced accidental deaths assumed to be caused by ordinary rearing environmental factors, which have not been experienced with existing immunodeficient animals. These incidents confirmed that NOG mice are completely different from existing immunodeficient mice and indicated that a higher-level infection control system than that of the standard rearing environment and an environmental control system that can reduce stress are indispensable in rearing these mice and in experiments using them. First some of these incidents are explained and then the necessity of a higher level of environmental control in animal experiments using NOG mice is proposed based on the knowledge gained in determining the causes of these deaths.

4.3.2 Opportunistic Infections

We experienced deaths due to *P. aeruginosa* in two research institutes performing human tumor and cord blood transplantation experiments on NOG mice. The deaths occurred from 1 to 1.5 years after the start of the experiments in both facilities, and debilitated and dead animals were found in both the experimental and nonexperimental groups. Since these mice were found to have interstitial pneumonia, suppurative nephritis, or hepatitis on necropsy (Fig. 7) and *P. aeruginosa* was isolated from the blood and lesions, the cause of death was diagnosed as bacteremia caused by *P. aeruginosa*. These incidents appeared to be due to contamination by *P. aeruginosa* of experimental materials such as the human tumors used in transplantation that spread to the rearing environment via the drinking water, researchers, or animal caretakers. A decrease in the cleanliness of the rearing facilities was also considered to be a cause of the spread of contamination. To prevent recurrences, the facilities are disinfected periodically and a system of microbiological testing of experimental materials is established so that this will not happen again.

These incidents led to concern about opportunistic pathogens such as *P. aeruginosa* that are not pathogenic in ordinary animals but are strongly pathogenic in NOG mice. The results of infection experiments performed to compare sensitivity of NOG and other immunodeficient mice to *P. aeruginosa* and *P. pneumotropica*, pathogens of opportunistic infections, are briefly described as follows. Nude and scid mice showed resistance to *P. aeruginosa*, but NOD-*scid* mice died of bacteremia at 2 weeks after administration, while NOG mice died of bacteremia at 1 week after administration. NOG mice also showed higher sensitivity to *P. pneumotropica* than other immunodeficient mice and died of pneumonia only 1 week after administration. These experimental results showed that in experiments using NOG mice

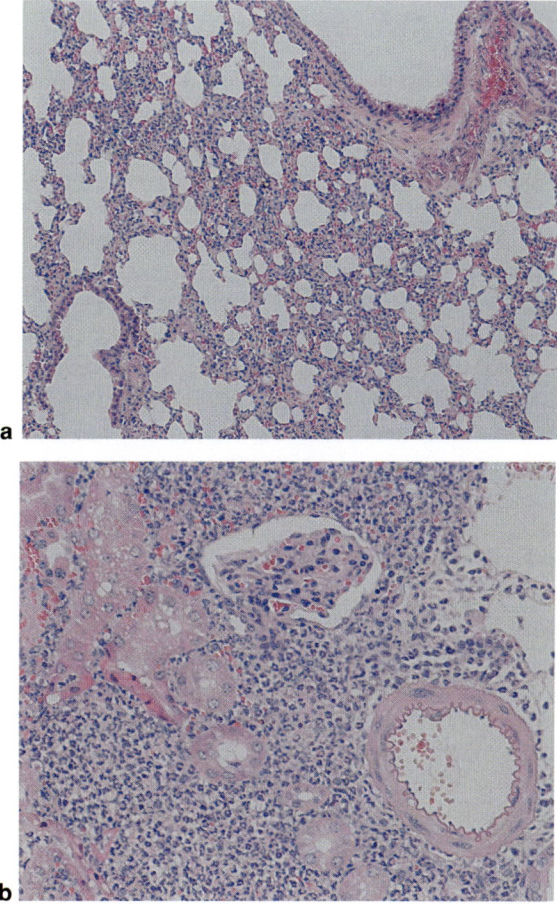

Fig. 7 Histopathological diagnosis of bacteremia in NOG mice. Interstitial pneumonia and suppurative nephritis were found in the lungs and kidneys of affected mice. **a** Interstitial pneumonitis (H&E, ×200). **b** Suppurative nephritis (H&E, ×200)

measures to protect against infections by opportunistic pathogens such as *P. aeruginosa* and *P. pneumotropica*, as well as *S. aureus* and *P. carinii*, are indispensable. These incidents suggested the importance of microbiological testing of transplantation materials such as tumors and cord blood.

4.3.3 Intestinal Flora and Human Disease Models

Differences and changes in the intestinal flora, which are the basis of the physiological functions of animals, cause diseases such as diarrhea and also affect responses to drugs, reproducibility of animal experiments, and expression of the characteristics of human disease models [13, 18, 22]. The effects on expression of the characteristics of human disease models include effects of different intestinal flora on carcinogenicity rates in *c-Ha-ras* transgenic mice [24] and effects due to changes in segmented filamentous bacteria that form the intestinal flora characteristic of the AID IgA deficient animal model. Such intestinal flora is known to affect the dramatype of human disease models [8, 35]. In NOG mice, the intestinal flora is susceptible to changes in environmental factors, and this can cause deaths in our experience. The intestinal flora of NOG mice has been found to change because of stress caused by transport or changes in rearing temperature, including cases where the *Escherichia coli* count increased because of transport-related stress and where *E. coli* increased in NOG mice reared in a high-temperature environment and did not recover thereafter. Therefore, environmental control to stabilize intestinal flora of NOG mice and establishment of a system of monitoring of such control are necessary for conducting highly safe and reproducible experiments using NOG mice.

4.3.4 Rearing Environment Stress

These problems arose in two research facilities (facilities A and B) conducting human bone marrow cell transplantation experiments using NOG mice. The time of onset was 1 to 1.5 years after the start of the experiments as in the incidents described above. In the beginning, it was assumed to be an infection and various microbiological tests were conducted, but no pathogens were detected in the affected animals. Because of the clinical symptoms and pathological findings, it was evident that infection was not the cause and the possibility of stress due to rearing environmental factors was considered.

In facility A, debility and mild diarrhea were observed in affected animals and sporadic deaths occurred. Necropsy and histopathological findings included duodenitis and enteritis associated with hyperplasia of the mucosal epithelium. Characteristic changes in the intestinal flora were observed, and some animals had no *E. coli*, which is normally detected in NOG mice.

In facility B, body weight decrease, severe diarrhea, and debility were observed in affected animals, and more than 100 NOG mice died. Necropsy findings included cholestasis, duodenitis, and enteritis (Fig. 8), and the histopathological findings included desquamation of the jejunal mucosal epithelium (Fig. 9). Changes in the intestinal flora were also observed, but unlike in facility A, the *E. coli* count, which is normally 10^5 -10^6 CFU/ml, increased to about 10^{10} CFU/ml.

Various stress factors were found in the rearing and experimental environments in both facilities: The animal rooms were not disinfected regularly, many nondesignated researchers had access to the experimental facilities, NOG mice were reared in the same room with other strains of mice, experimental treatment was performed in different rooms, and animal experiments were conducted at night. These multiple stresses appeared to destroy the physiological balance of the NOG mice, resulting in debility and death. As measures to prevent recurrences, the

Basic Concept of Development and Practical Application

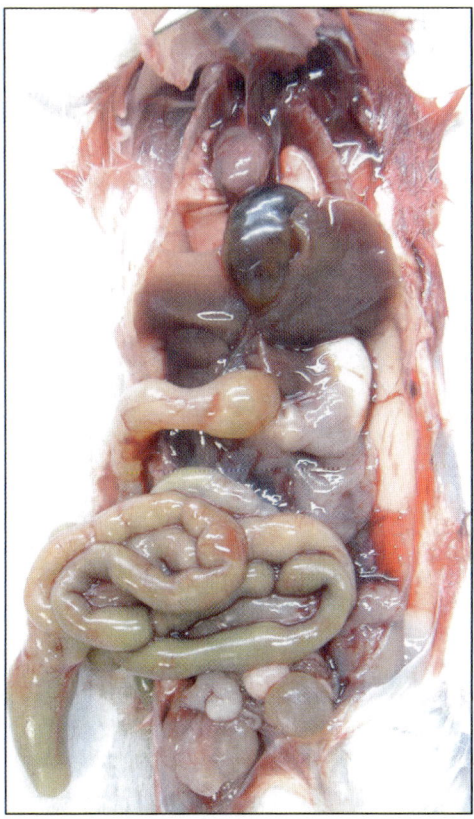

Fig. 8 Results of necropsy of diarrheal NOG mice. Severe diarrhea, bile congestion, duodenitis, and intestinal hypertrophy were observed

animal rooms were regularly disinfected, the NOG were reared separately, access of personnel was restricted, moving the animals for experimental treatment was prohibited, and performing experiments at night was also prohibited. Presumably as a result, no recurrences have occurred.

4.3.5 Rearing Environment for Animal Experiments Using NOG Mice

The effects of *P. aeruginosa* infection on animal experiments have been reported, including bacteremia caused by experimental treatment such as irradiation [9, 10], but there have been no reports of deaths due to spontaneous infections in existing immunodeficient animals. Therefore, the deaths caused by *P. aeruginosa* infection introduced here were due to the fact that the mice involved were NOG mice that are severely immunodeficient, which suggests that elimination of opportunistic pathogens such as *P. aeruginosa* from the rearing environment is essential.

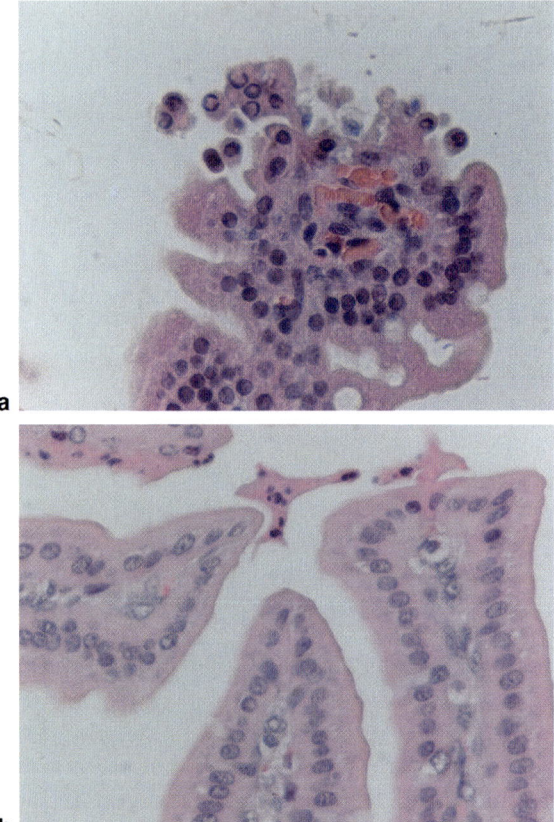

Fig. 9 Histopathological diagnosis of intestine in diarrheal NOG mice (H&E, ×200) A decidual mucosal epithelium was found in the jejunum. **a** Jejunum of diarrheal NOG mice. **b** Jejunum of normal NOG mice

The cause of stress-related deaths is considered to be stress caused by the rearing environment in which the animal experiments are performed, which usually presents no problem. This shows the necessity of reducing stress in the environment of animal experiments using NOG mice.

These incidents confirmed that NOG mice have completely different physiological functions from existing immunodeficient animals. The following rearing environmental conditions are minimum requirements for safe animal experiments using NOG mice.

1. Establishment of a strict infection defense system (especially preventive measures against opportunistic infections)
2. Establishment of a microbiological monitoring system including opportunistic pathogens for immunodeficient animals

3. Establishment of a microbiological testing system for experimental materials such as tumors
4. Establishment of a highly clean rearing environment
5. Reduction of stress caused by environmental factors

Since the incidents described above, similar sporadic deaths have occurred in animal experimentation facilities using NOG mice. We intend to analyze the causes of these incidents and publish the information obtained. We also plan to collect background data of NOG mice, construct a system for monitoring changes in physiological functions using intestinal flora as the parameter, and establish a environmental control system for animal experiments so that animal experiments using NOG mice as highly useful human disease models can be performed safety with a high level of reproducibility.

5 Future Aspects of Humanized Mice

A mouse with a functioning human organ system has long been one of the major targets in biomedical research. The first reported case was the SCID-hu model reconstituted by human fetal liver and thymus in *scid* mice [20]. The function of the human immune system reconstituted in immunodeficient mice lacking specific immunity such as scid mice was reported to be very limited. Recently, it became evident that IL-2 receptor common γ chain deficiency combined with NOD-*scid* or Rag2 knockout mice created an extremely immunodeficient phenotype that enabled not only permanent engraftment but also differentiation and proliferation of the human progeny of hematopoietic/immunocyte system. It was also shown that these mice should be very useful for biomedical research in the fields of stem cell biology, infectious disease, or regenerative therapy for a variety of diseases. These results promise an enormous contribution to human health and welfare in the future. We hope to see further developments in bioscience with these excellent tools.

6 Conclusion

This review describes a basic concept for development of animal models in CIEA, which show, in part, biological functions identical to those found in human beings. Standardized laboratory animals used for biomedical research are mass produced to comply with the requirements set in the standards or specifications. Therefore, quality standards for genetic and microbiological factors are absolutely essential for a standardized laboratory animal. The genetic quality determines the "phenotype," and the phenotype and proximate environmental factors determine the "dramatype." Because the results of animal experiments are the response of the dramatype to the experimental procedure, not only strict management of genetic quality but also proximate environmental control is needed to ensure reproducibility of experimental results. The NOG mice supplied at present comply with these standards, and it is possible to perform experiments using NOG mice with the same genetic quality anywhere in the world. Environmental control is complicated and more difficult than genetic quality control, but we have clarified the minimum requirements for rearing environmental conditions for stable animal experiments using NOG mice through experience over the past 4 years. We are also attempting to establish an environmental control system for animal experiments so that experiments using NOG mice can be performed safely with a high level of reproducibility. It must be kept in mind that NOG mice are known to have completely different resistance to stress and infections from conventional immunodeficient mice. It is also indispensable to control environmental factors, including microbiological factors, which have critical effects on the results of animal experiments.

References

1. Bosma, G. C., R. P. Custer, and M. J. Bosma. 1983. A severe combined immunodeficiency mutation in the mouse. Nature 301:527-530.
2. Brooks, C. G., P. J. Webb, R. A. Robins, G. Robinson, R. W. Baldwin, and M. F. Festing. 1980. Studies on the immunobiology of rnu/rnu "nude" rats with congenital aplasia of the thymus. Eur J Immunol 10:58-65.
3. Diwan, B. A., J. M. Rice, M. Ohshima, and J. M. Ward. 1986. Interstrain differences in susceptibility to liver carcinogenesis initiated by N-nitrosodiethylamine and its promotion by phenobarbital in C57BL/6NCr, C3H/HeNCrMTV- and DBA/2NCr mice. Carcinogenesis 7:215-220.
4. Dragunsky, E., T. Nomura, K. Karpinski, J. Furesz, D. J. Wood, Y. Pervikov, S. Abe, T. Kurata, O. Vanloocke, G. Karganova, R. Taffs, A. Heath, A. Ivshina, and I. Levenbook. 2003. Transgenic mice as an alternative to monkeys for neurovirulence testing of live oral poliovirus vaccine: validation by a WHO collaborative study. Bull World Health Organ 81:251-260.
5. Dragunsky, E., R. Taffs, Y. Chernokhvostova, T. Nomura, K. Hioki, D. Gardner, L. Norwood, and I. Levenbook. 1996. A poliovirus-susceptible transgenic mouse model as a possible replacement for the monkey neurovirulence test of oral poliovirus vaccine. Biologicals 24:77-86.
6. Drinkwater, N. R., and J. J. Ginsler. 1986. Genetic control of hepatocarcinogenesis in C57BL/6J and C3H/HeJ inbred mice. Carcinogenesis 7:1701-1707.
7. Estill, S. J., and J. A. Garcia. 2000. A marker assisted selection protocol (MASP) to generate C57BL/6J or 129S6/SvEvTac speed congenic or consomic strains. Genesis 28:164-166.
8. Fagarasan, S., M. Muramatsu, K. Suzuki, H. Nagaoka, H. Hiai, and T. Honjo. 2002. Critical roles of activation-induced cytidine deaminase in the homeostasis of gut flora. Science 298:1424-1427.
9. Flynn, R. J. 1963. The diagnosis of *Pseudomonas aeruginosa* infection of mice. Lab Anim Care 13:SUPPL126-129.
10. Hightower, D., H. T. Uhrig, and J. I. Davis. 1966. *Pseudomonas aeruginosa* infection in rats used in radiobiology research. Lab Anim Care 16:85-92.
11. Horie, H., S. Koike, T. Kurata, Y. Sato-Yoshida, I. Ise, Y. Ota, S. Abe, K. Hioki, H. Kato, C. Taya et al. 1994. Transgenic mice carrying the human poliovirus receptor: new animal models for study of poliovirus neurovirulence. J Virol 68:681-688.

12. Ito, M., H. Hiramatsu, K. Kobayashi, K. Suzue, M. Kawahata, K. Hioki, Y. Ueyama, Y. Koyanagi, K. Sugamura, K. Tsuji, T. Heike, and T. Nakahata. 2002. NOD/SCID/γ_c^{null} mouse: an excellent recipient mouse model for engraftment of human cells. Blood 100:3175-3182.
13. Itoh, K., T. Mitsuoka, K. Sudo, and K. Suzuki. 1983. Comparison of fecal lactobacilli in mice of different strains under different housing conditions. Z Versuchstierkd 25:193-200.
14. Koike, S., H. Horie, Y. Sato, I. Ise, C. Taya, T. Nomura, I. Yoshioka, H. Yonekawa, and A. Nomoto. 1993. Poliovirus-sensitive transgenic mice as a new animal model. Dev Biol Stand 78:101-107.
15. Koyanagi, Y., Y. Tanaka, J. Kira, M. Ito, K. Hioki, N. Misawa, Y. Kawano, K. Yamasaki, R. Tanaka, Y. Suzuki, Y. Ueyama, E. Terada, T. Tanaka, M. Miyasaka, T. Kobayashi, Y. Kumazawa, and N. Yamamoto. 1997. Primary human immunodeficiency virus type 1 viremia and central nervous system invasion in a novel hu-PBL-immunodeficient mouse strain. J Virol 71:2417-2424.
16. Lee, G. H., K. Nomura, H. Kanda, M. Kusakabe, A. Yoshiki, T. Sakakura, and T. Kitagawa. 1991. Strain specific sensitivity to diethylnitrosamine-induced carcinogenesis is maintained in hepatocytes of C3H/HeN in equilibrium with C57BL/6N chimeric mice. Cancer Res 51:3257-3260.
17. Levenbook, I., and T. Nomura. 1997. Development of a neurovirulent testing system for oral poliovirus vaccine with transgenic mice. Lab Anim Sci 47:118-120.
18. Mastromarino, A., B. S. Reddy, and E. L. Wynder. 1976. Metabolic epidemiology of colon cancer: enzymic activity of fecal flora. Am J Clin Nutr 29:1455-1460.
19. Masuda, H., T. Maruyama, E. Hiratsu, J. Yamane, A. Iwanami, T. Nagashima, M. Ono, H. Miyoshi, H. J. Okano, M. Ito, N. Tamaoki, T. Nomura, H. Okano, Y. Matsuzaki, and Y. Yoshimura. 2007. Noninvasive and real-time assessment of reconstructed functional human endometrium in NOD/SCID/gamma c^{null} immunodeficient mice. Proc Natl Acad Sci USA 104:1925-1930.
20. McCune, J. M., R. Namikawa, H. Kaneshima, L. D. Shultz, M. Lieberman, and I. L. Weissman. 1988. The SCID-hu mouse: murine model for the analysis of human hematolymphoid differentiation and function. Science 241:1632-1639.
21. Mitsumori, K., S. Wakana, S. Yamamoto, Y. Kodama, K. Yasuhara, T. Nomura, Y. Hayashi, and R. R. Maronpot. 1997. Susceptibility of transgenic mice carrying human prototype c-Ha-ras gene in a short-term carcinogenicity study of vinyl carbamate and ras gene analyses of the induced tumors. Mol Carcinog 20:298-307.
22. Mizutani, T., and T. Mitsuoka. 1980. Inhibitory effect of some intestinal bacteria on liver tumorigenesis in gnotobiotic C3H/He male mice. Cancer Lett 11:89-95.
23. Nakanishi, T., A. Kuroiwa, S. Yamada, A. Isotani, A. Yamashita, A. Tairaka, T. Hayashi, T. Takagi, M. Ikawa, Y. Matsuda, and M. Okabe. 2002. FISH analysis of 142 EGFP transgene integration sites into the mouse genome. Genomics 80:564-574.
24. Narushima, S., K. Itoh, T. Mitsuoka, H. Nakayama, T. Itoh, K. Hioki, and T. Nomura. 1998. Effect of mouse intestinal bacteria on incidence of colorectal tumors induced by 1,2-dimethylhydrazine injection in gnotobiotic transgenic mice harboring human prototype c-Ha-ras genes. Exp Anim 47:111-117.
25. Nomura, T. 1997. Practical development of genetically engineered animals as human disease models. Lab Anim Sci 47:113-117.
26. Nomura, T., K. Esaki, and T. Tomita. 1984. ICLAS Manual for Genetic Monitoring of Inbred Mice, University of Tokyo Press, Tokyo, Japan.
27. Nomura, T., and C. E. Hopla. 1985. ICLAS Reference and Monitoring Centers Program. 8th ICLAS/CALAS Symp., Vancouver 1983. Gustav Fischer Verlag, Stuttgart, New York, 1221-1224.
28. Nomura, T., and N. Kagiyama. 1982. Importance of microbiological control in using nude mice. The 3rd International Workshop on Nude Mice, 11-21.
29. Nomura, T., and C. Yamauchi. 1968. Environments and Physiological Status of Experimental Animals. Experimental Animals In Cancer Res. 17-35.
30. Nomura, T., C. Yamauchi, and H. Takahashi. 1967. Influence of environmental temperature on physiological functions of the laboratory mouse. Husbandry of laboratory animals.

459-470 (Proceedings of the 453rd Intern. Symp. organized by the Intern. Comm. on Laboratory Animals,1965, Ireland).
31. Peter, D. W., and D. P. J. Raloh. 1982. Experimental *Pneumocystis carinii* infection in nude and steroid-treated normal mice. The 3rd International Workshop on Nude Mice:11-21.
32. Russell, W. M. S., and R. L. Burch. 1959. The principals of humane experimental technique. 238 pp, Methuen, Bethesda.
33. Rygaard, J., and C. O. Povlsen. 1969. Heterotransplantation of a human malignant tumour to "Nude" mice. Acta Pathol Microbiol Scand 77:758-760.
34. Shultz, L. D., P. A. Schweitzer, S. W. Christianson, B. Gott, I. B. Schweitzer, B. Tennent, S. McKenna, L. Mobraaten, T. V. Rajan, D. L. Greiner et al. 1995. Multiple defects in innate and adaptive immunologic function in NOD/LtSz-scid mice. J Immunol 154:180-191.
35. Suzuki, K., B. Meek, Y. Doi, M. Muramatsu, T. Chiba, T. Honjo, and S. Fagarasan. 2004. Aberrant expansion of segmented filamentous bacteria in IgA-deficient gut. Proc Natl Acad Sci USA 101:1981-1986.
36. Tanioka, Y., K. Esaki, C. Yamauchi, and T. Nomura. 1968. Influence of environmental temperature on the mortality of anaphylactic shock in mice. Exp Animals 17:7-10.
37. Tanioka, Y., K. Esaki, C. Yamauchi, and T. Nomura. 1971. Effect of the change in environmental temperature on the incidence of malformations in the mouse fetuses induced by ethylurethane injection. Cong Anom 11:19-23.
38. Thompson, K. L., B. A. Rosenzweig, R. Honchel, R. E. Cannon, K. T. Blanchard, R. E. Stoll, and F. D. Sistare. 2001. Loss of critical palindromic transgene promoter sequence in chemically induced Tg.AC mouse skin papillomas expressing transgene-derived mRNA. Mol Carcinog 32:176-186.
39. Waggie, K., N. Kagiyama, A. M. Allen, and T. Nomura. 1994. Manual of Microbiologic Monitoring of Laboratory Animals. National Institute of Health (NIH Publication No. 94-2498), Bethesda.
40. Wong, G. T. 2002. Speed congenics: applications for transgenic and knock-out mouse strains. Neuropeptides 36:230-236.
41. Yamamoto, S., Y. Hayashi, K. Mitsumori, and T. Nomura. 1997. Rapid carcinogenicity testing system with transgenic mice harboring human prototype c-HRAS gene. Lab Anim Sci 47:121-126.
42. Zumbach, A., G. A. Marbet, and D. A. Tsakiris. 2001. Influence of the genetic background on platelet function, microparticle and thrombin generation in the common laboratory mouse. Platelets 12:496-502.

Humanized SCID Mouse Models for Biomedical Research

T. Pearson, D. L. Greiner, and L. D. Shultz(✉)

1	Introduction	26
	1.1 Need for Small Animal Models for Preclinical Research	26
	1.2 Humanized Mice	27
	1.3 Historical Perspective	27
2	Manipulations Used to Improve Human Hematolymphoid Cell Engraftment in Immunodeficient Mice	29
	2.1 Manipulations to Depress Innate Immunity	29
	2.2 Exogenous Administration of Human Hormones, Growth Factors, and Cytokines	30
3	Transgenic Expression of Human Hormones, Growth Factors, Cytokines and HLA Molecules	32
	3.1 Transgenic Expression of Human Hormones, Growth Factors, and Cytokines	32
	3.2 Transgenic Expression of Human HLA Molecules	32
4	Ex Vivo Modifications of Human HSC Before Engraftment	33
5	Coinjection of HSC with Mesenchymal Stem Cells and Other Stromal Cell Populations	35
6	The IL2 Receptor Gamma Common Chain Knockout Mouse	35
7	Current Research Uses of Humanized Mice	37
	7.1 Infectious Disease	37
	7.2 Autoimmunity	37
	7.3 Cancer	38
	7.4 Regenerative Medicine	39
8	Remaining Hurdles and Future Directions	40
	8.1 Human-Specific Molecule Expression	40
	8.2 New Approaches for Reducing Residual Innate Immunity	41
	8.3 Artificial Organoids	41
9	Conclusions	42
	References	42

L. D. Shultz
The Jackson Laboratory, 600 Main Street, Bar Harbor, ME 04609, USA
lenny.Shultz@jax.org

Abstract There is a growing need for effective animal models to carry out experimental studies on human hematopoietic and immune systems without putting individuals at risk. Progress in development of small animal models for the in vivo investigation of human hematopoiesis and immunity has seen three major breakthroughs over the last three decades. First, CB17-*Prkdc*^{scid} (abbreviated CB17-*scid*) mice were discovered in 1983, and engraftment of these mice with human fetal tissues (SCID-Hu model) and peripheral blood mononuclear cells (Hu-PBL-SCID model) was reported in 1988. Second, NOD-*scid* mice were developed and their enhanced ability to engraft with human hematolymphoid tissues as compared with CB17-*scid* mice was reported in 1995. NOD-*scid* mice have been the "gold standard" for studies of human hematolymphoid engraftment in small animal models over the last 10 years. Third, immunodeficient mice bearing a targeted mutation in the IL-2 receptor common gamma chain (*IL2rγ*^{null}) were developed independently by four groups between 2002 and 2005, and a major increase in the engraftment and function of human hematolymphoid cells as compared with NOD-*scid* mice has been reported. These new strains of immunodeficient *IL2rγ*^{null} mice are now being used for studies in human hematopoiesis, innate and adaptive immunity, autoimmunity, infectious diseases, cancer biology, and regenerative medicine. In this chapter, we discuss the current state of development of these strains of mice, the remaining deficiencies, and how approaches used to increase the engraftment and function of human hematolymphoid cells in CB17-*scid* mice and in previous models based on NOD-*scid* mice may enhance human hematolymphoid engraftment and function in NOD-*scid IL2rγ*^{null} mice.

Abbreviations AML: acute myelogenous leukemia; asialoGM1: asialoganglioside ganliotetraosylceramide; BAFF: B cell activating factor; Blys: B lymphocyte stimulator factor; BM: bone marrow; DTR: diphtheria toxin receptor; ES: embryonic stem; FLT-3L: fms-related tyrosine kinase 3 ligand; EPO: erythropoietin; GM-CSF: granulocyte-macrophage colony-stimulating factor; HSC: hematopoietic stem cell; IL: interleukin; IC: intracardiac; IP: intraperitoneal; IV: intravenous; mAb: monoclonal antibody; MGDF: megakaryocyte growth and development factor; MHC: major histocompatibility complex; MSC: mesenchymal stem cell; NK: natural killer; PBMC: peripheral blood mononuclear cells; RBC: red blood cells; SCF: stem cell factor; SDF-1: stromal cell-derived factor-1; TCR: T cell receptor; TNFα: tumor necrosis factor-α; TPO: thrombopoietin; USSC: unrestricted somatic stem cells; UCB: umbilical cord blood

1 Introduction

1.1 Need for Small Animal Models for Preclinical Research

Animal models, particularly mice and rats, have provided important fundamental insights into biological and immunological processes that are common between species. However, direct translation of these results from rodents to humans often fails because of species-specific differences. Furthermore, in vivo experimentation on humans is constrained by ethical and technical concerns. There is a critical need to develop animal models for the study of human immunity and other complex human biological processes without putting individuals at risk. Many of these processes involve complex biological systems that cannot be modeled in vitro or ex vivo. Furthermore, our understanding of biological process in humans is not sufficient to permit in silico modeling of the complex traits. Based on the discovery of immunodeficient CB17-*scid* mice in 1983, investigators have attempted to "humanize" mice to model complex human immunological and biological processes in small animals. Improvements in the generation of humanized mice have led to their growing use in the investigation of human hematopoiesis, innate and adaptive immunity, autoimmunity, infectious diseases, cancer biology, and, more recently, regenerative medicine.

1.2 Humanized Mice

Transplantation of human cells or tissues and transgenic expression of human molecules such as major histocompatibility complex (HLA) antigens all fall under the umbrella of the generic term "humanized" mice. Humanization by transgene expression can be used in immunocompetent mice to identify, for example, the antigenic epitopes presented by a human HLA molecule to an immune system [120]. Alternatively, engraftment of immunodeficient mice with human hematopoietic stem cells (HSC) or human peripheral blood mononuclear cells (PBMC) permits investigation of the development and function of a human immune system in vivo. We first provide a historical perspective on the development of humanized mice, approaches that have been used to enhance the engraftment and function of the transplanted human cells and tissues, and the exciting opportunities that are now possible based on the development of immunodeficient mice with targeted mutations in the IL-2 receptor common gamma chain.

1.3 Historical Perspective

Initial attempts to engraft human hematolymphoid cells in *Foxn1^{nu}* (abbreviated nude) mice were disappointing, even when the *Lyst^{bg}* (beige) and *Btk^{xid}* (xid) mutations were crossed onto nude mice [25]. The first major breakthrough in the ability of mice to be engrafted with human hematolymphoid cells was the discovery of CB17-*scid* mice in 1983 [14]. This was followed by the demonstration in 1988 that human fetal tissues and PBMC could engraft in CB17-*scid* mice [71, 74] (Fig. 1). Limitations of this model included high levels of host NK cell activity and the development of murine T and B cells on aging ("leakiness"), resulting in their ability to support only very low levels of human hematolymphoid engraftment [41].

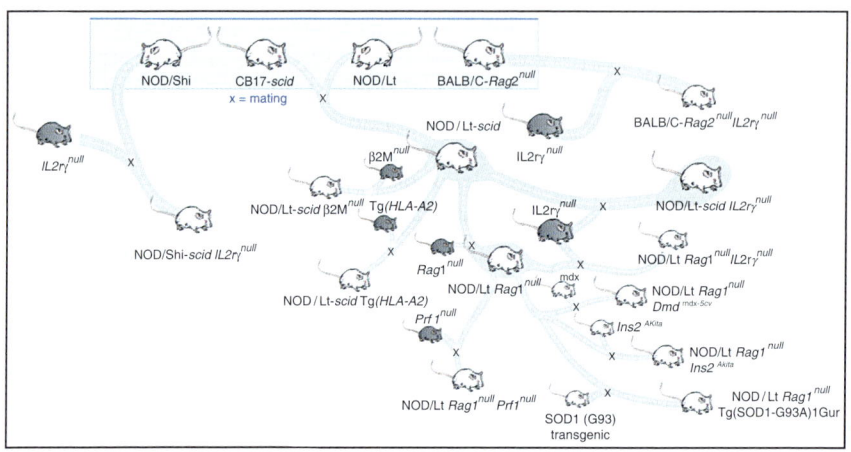

Fig. 1 A road map showing many of the diverse immunodeficient mouse stocks used for engraftment of human cells and tissues. The generation of these stocks of immunodeficient mice for humanization was based on the discovery of the *scid* mutation in 1983. The second major breakthrough was based on the development of the NOD-*scid* strain of mice. The most recent breakthrough in humanization is based on the generation of immunodeficient mice bearing a targeted mutation in the IL-2 receptor gamma chain gene. At the Jackson Laboratory, we have used the NOD-*scid* and NOD-*Rag1*^{null} strains as the base stocks of mice for generating immunodeficient *IL2rγ*^{null} mice and all future genetic modifications of these mice. These modifications leading to improvements in humanization of the model system will be useful for targeted research applications for diseases such as diabetes, muscular dystrophy, and various neurological disorders

Attempts to reduce innate immunity and increase the engraftment and function of human hematolymphoid cells included backcrossing the *scid* mutation onto other strains of mice with defects in innate immunity such as C3H/HeJ mice that express macrophage abnormalities or mice bearing the beige mutation that leads to defects in NK cell cytotoxicity [44]. However, these genetic manipulations led to only incremental improvements in human hematolymphoid cell engraftment [110].

The second major breakthrough was the development of NOD/LtSz-*Prkdc*^{scid} (abbreviated NOD-*scid*) mice in 1995 (Fig. 1). NOD-*scid* mice have reduced levels of NK cell activity and additional deficiencies in innate immunity [107] and support heightened levels of human hematolymphoid cell engraftment as compared to CB17-*scid* mice [44, 65, 86]. Since its development in 1995, this model has undergone many genetic modifications in attempts to improve the engraftment and function of human hematolymphoid cells and tissues. These modifications included efforts to further decrease innate immunity by generating NOD-*scid* mice homozygous for a targeted mutation at the β2-microglobulin (*B2m*) locus that results in NK cell deficiency [23] or the null mutation in the perforin (*Prf1*) locus) that markedly reduces NK cell-mediated cytotoxicity [109]. Additionally, a more radioresistant immunodeficient stock of NOD mice was generated by introducing a null mutation in the recombination activating gene 1 (*Rag1*) locus [108]. Finally, mice expressing human HLA molecules and cytokines via transgenesis have been generated. Together, these genetic modifications improved human lymphohematopoietic cell engraftment over CB17-*scid* mice, but models based on NOD-*scid* mice remained limited by their short life span due to the early development of thymic lymphomas and the lack of development of a fully functional human immune system after human HSC engraftment.

The third and most recent breakthrough in the field was the development of immunodeficient mice carrying a targeted mutation at the interleukin 2 receptor common gamma (*Il2rg*) chain (hereafter abbreviated *IL2rγ^{null}*) (Fig. 1). Deficiency of the *Il2rg* chain causes X-linked SCID in humans [60]. This molecule is utilized by a number of cytokine receptors and is indispensable for IL-2, IL-4, IL-7, IL-9, IL-15, and IL-21 high-affinity ligand binding and signaling [118]. Thus, mice bearing an *IL2rγ^{null}* mutation have severe impairments in innate and adaptive immunity. Mice harboring an *IL2rγ^{null}* mutation in combination with the *scid*, *Rag1^{null}*, or *Rag2^{null}* mutations show a major increase in their ability to support engraftment of functional human hematolymphoid cells.

IL2rγ targeted mutations have been produced independently by four different groups [17, 29, 51, 82], and these *IL2rγ^{null}* genetic stocks including *Il2rg^{tm1Wjj}*, *Il2rg^{tm1Sug}*, and *Il2rg^{tm1Krf}*, and *Il2rg^{tm1Cgn}* have been bred to *scid*, *Rag1^{tm1Mom}* (*Rag1^{null}*), or *Rag2^{tm1Fwa}* (*Rag2^{null}*) mice to develop models for human hematolymphoid engraftment [110]. However, there remain limitations on the engraftment and function of human hematolymphoid cells even in these new immunodeficient *IL2rγ^{null}* genetic stocks. Lessons learned over the last 10 years from approaches used to enhance the ability of previous generations of immunodeficient mice to engraft with human hematolymphoid cells may provide insights into approaches that will facilitate engraftment and function of human hematolymphoid cells in NOD-*scid IL2rγ^{null}* mice.

2 Manipulations Used to Improve Human Hematolymphoid Cell Engraftment in Immunodeficient Mice

2.1 Manipulations to Depress Innate Immunity

Mice bearing the *scid* mutation or the *Rag1^{null}* or *Rag2^{null}* targeted mutations lack adaptive immunity. However, depending on the strain background, these immunodeficient mice retain robust innate immunity. This innate immunity poses a significant barrier to the engraftment of xenogeneic hematopoietic tissues [22]. Beginning with the CB17-*scid* strain, which expressed high levels of host innate immune activity, many methods have been used to depress innate immunity to increase human hematolymphoid cell engraftment (Table 1). Initially, these reagents targeted NK cells, which are known to be a major obstacle to engraftment of hematopoietic cells [42]. The original approach used antibody against asialoganglioside ganli-

Table 1 Experimental approaches used to decrease innate immunity in SCID mice

Target Cells	Treatment	Reference
NK cells	Anti–asialo–GM–1 antibody	99, 105
NK cells	Anti–CD122 (TMβ1) mAb	8, 72, 121
NK cells	Anti–NK1.1 (PK-136) mAb	22
Granulocytes	Anti–GR1 (RB6–8C5) mAb	100
Macrophages	Liposome–encapsulated clodronate	34, 97, 104, 125, 131

otetraosylceramide (asialoGM1), but this antibody is highly cross-reactive between species and also targets human NK cells, activated CD8$^+$ T cells [119], and macrophages [73] that are present in the human cell inoculum. In strains such as C57BL/6-*scid* mice that express the NK1.1 allele, anti-NK1.1 monoclonal antibody (mAb) has been used to deplete NK cells [22]. However, even in anti-NK1.1 mAb-treated C57BL/6 mice, human hematolymphoid cell engraftment remained low [22]. Interestingly, this also demonstrates the importance of mouse strain background in the support of human cell engraftment in immunodeficient hosts. The more recent use of anti-CD122 (IL-2 receptor beta chain) mAb, which targets a molecule that is highly expressed on murine NK cells and does not cross-react with human CD122 has facilitated human hematopoietic cell engraftment in NOD-*scid* mice. Finally, other approaches have targeted neutrophils with anti-Gr1 mAb, and macrophages have been depleted with liposome-encapsulated clodronate (Table 1).

2.2 *Exogenous Administration of Human Hormones, Growth Factors, and Cytokines*

An additional obstacle to human cell engraftment and function, recognized early in the course of investigations using humanized mice, is the lack of species cross-reactivity of many mouse hormones, growth factors, and cytokines that are required for development, survival, and function of human hematolymphoid cells (http://www.copewithcytokines.de/cope.cgi?key=Cytokine%20Inter-species%20Reactivities). For example, mouse and human type 1 interferon are species specific and do not cross-react [128]. It has also been reported that granulocyte monocyte colony-stimulating factor (GM-CSF) and IL-3 are species specific and are not cross-reactive between mice and humans, meaning that these mouse cytokines will not support the growth of engrafted human cells [5]. In contrast, factors such as erythropoietin (EPO), granulocyte colony-stimulating factor (G-CSF), and stem cell factor (SCF or kit-ligand) are cross-reactive between human and mouse, and host production of these molecules can support human cell growth [5]. In addition, many necessary human-specific factors are not produced by the human hematolymphoid cells, but rather by nonhematopoietic stromal cells not present in the human HSC inoculum. An example of this is the B cell cytokine B lymphocyte stimulator (BLyS, also termed B cell activating factor or BAFF, official gene nomenclature, *TNFSF13B*) [136]. BLyS, produced by stromal cells and dendritic cells, is required for B cell differentiation and survival. Our collaborators (Drs. Woodland and Schmidt at the University of Massachusetts) have found that murine BLyS fails to promote human B cell survival ex vivo or in humanized mice. These observations suggest that human BLyS may need to be provided, either by exogenous administration or by transgenic expression, for robust B cell engraftment and function in humanized mice.

In an attempt to overcome these obstacles, investigators have provided exogenous human factors to human hematolymphoid cell-engrafted mice. These factors comprise three main categories. Those important for (1) the engraftment of human HSC, (2) the differentiation of human HSC into various hematopoietic lineages, and (3) the function of the differentiated cells in the immunodeficient murine host. Exogenous administration of cytokines and factors required for early HSC homing to the bone marrow and survival have been identified and used to improve engraftment of human HSC. These factors targeted homing molecules such as stromal cell-derived factor-1 (SDF-1) and molecules required for HSC expansion and differentiation such as IL-3, FMS-related tyrosine kinase 3 ligand (FLT-3 ligand), and GM-CSF (Table 2).

A second series of exogenous human factors administered to engrafted mice included those targeted toward differentiation and function of specific human hematopoietic lineages. The inability of NOD-*scid* mice engrafted with human HSC to generate human T cells prompted attempts to enhance T cell differentiation. Important advances in this area include the administration of TNF-α [98] or IL-7 to

Table 2 Treatments used to promote human hematolymphoid cell survival, function, or trafficking in SCID mice

Treatment	Reference
Human growth hormone	75, 123
SDF–1 peptide agonist	84
Flt3 ligand	19, 53
Erythropoietin	19, 53
NIP–004 human TPO receptor agonist	78
TNF–α	98, 132
IL–2	7, 96
IL–3, GM–CSF, PIXY 321	7, 19, 59
IL–4	18
IL–6	19, 53, 56
IL–7	53, 79, 106
IL–12	102, 135
IL–15	96
IL–18	102
Type 1 interferon	58, 101
Anti–CD40 mAb	76
Anti–CTLA–4 mAb	69

human HSC-engrafted mice [106]. Additional lineage-specific factors such as EPO for red blood cell (RBC) development have also been used to enhance human RBC generation. Exogenous administration of EPO may be needed to enhance human RBC development in HSC-engrafted immunodeficient mice, as reports suggest that engrafted HSC generate only low levels of human RBC [48]. Finally, additional cytokines required for differentiation and function of human lymphocytes have been used to drive HSC differentiation toward the lymphocyte lineage, as well as molecules that provide costimulatory signals to cells of the developing immune system (Table 2). A third series of factors includes those directed at regulating the function or maturation of the engrafted human hematolymphoid cells. These include factors such as thrombopoietin, anti-CD40 mAb, anti-CTLA-4 mAb, and cytokines such as IL-12, which is known to be important in cell immune function (Table 2).

3 Transgenic Expression of Human Hormones, Growth Factors, Cytokines and HLA Molecules

A potentially more efficient approach to providing human molecules required for human HSC engraftment and function is through transgenic expression of growth factors. In addition, other noncytokine human molecules that are important for human cell engraftment of lineage development can be introduced via transgenesis. Examples include human HLA class I and class II molecules, which are discussed next.

3.1 Transgenic Expression of Human Hormones, Growth Factors, and Cytokines

One of the first attempts to use transgenic immunodeficient mice to enhance engraftment of human hematolymphoid cells was the transgenic expression of IL-3, GM-CSF, and SCF in CB17-*scid* mice [11]. However, human HSC engraftment was not improved, and subsequently transgenic expression of these three factors was determined to be detrimental to the engraftment and differentiation of human HSC [80]. This effect was hypothesized to result from either the increased mobilization of stem cell progenitors into the blood, effectively preventing their seeding into the bone marrow, or the induced terminal differentiation of the stem cells in the presence of high levels of these cytokines.

3.2 Transgenic Expression of Human HLA Molecules

A second class of human transgenes that have been expressed in mice are HLA class I and class II molecules. The use of human HLA class I [32] and class II [35] immunocompetent transgenic mice for the study of immune function and

Table 3 Examples of transgenic expression of human HLA molecules and hematopoietic growth factors

Human transgene expression	Reference
Human transgenes in SCID mice	
HLA–A2	8
HLA–DR1	16
IL3, GM–CSF, SCF	11, 80
Human HLA transgenes in immunocompetent mice	
HLA–A24	39
HLA–B27	57, 124
HLA–DR3	67
HLA–DR4	122, 137
HLA–DQ8	64, 77, 91

autoimmunity has been extensively reviewed. In immunodeficient humanized mice, expression of HLA molecules on host thymic epithelium is required for appropriate thymic selection and antigen-specific restriction by human T cells. To address this issue, numerous transgenic mice expressing human HLA molecules have been created (Table 3). However, only a few of these transgenes have been backcrossed onto immunodeficient mouse stocks for use in human HSC engraftment experiments. In the earlier generations of non-HLA-transgenic immunodeficient mice that have been used in humanization experiments, variable levels of human T cell development have been observed [16]. The generation of NOD-*scid Il2r*γ^{null} mice expressing human HLA transgenes should now permit the development of HLA-restricted human T cells in these mice.

4 Ex Vivo Modifications of Human HSC Before Engraftment

An alternative approach to the treatment or modification of the immunodeficient host to improve human hematolymphoid cell engraftment and differentiation is to perform ex vivo manipulations of the human HSC inoculum (Table 4). Indeed, there has been intense interest in the ex vivo expansion of human HSC. This is due in part to the need for sufficient HSC numbers to achieve efficacy in the clinic for human stem cell engraftment from umbilical cord blood (UCB), based on a recommended dose of 2.0×10^7 UCB nucleated cells per kilogram of recipient body weight [13].

Most approaches have relied on manipulations of cultured HSC with cocktails containing multiple human cytokines. These approaches have been hampered by the difficulty in achieving ex vivo expansion of human $CD34^+$ stem cells without inducing their differentiation and consequent loss of in vivo stem and progenitor cell repopulating capacity [113].

Noncytokine approaches have also been used in ex vivo manipulations to modify human HSC to facilitate their in vivo engraftment. For example, surface fucosylation of CD34$^+$ UCB stem cells has been reported to generate selectin ligands that enhance the initial interactions with microvessels following injection [45, 138]. Selectins contribute to homing of adult CD34$^+$ cells, but their ligands are absent on a subset of UCB CD34$^+$ cells. This deficiency has been associated with reduced α1,3-fucosyltransferase expression and activity. Exogenous treatment of UCB CD34$^+$ stem cells to introduce α1,3-linked fucose to cell surface glycans on their surface has been used successfully to enhance their engraftment in vivo [45, 138].

A third approach has used lineage-specific differentiation cytokines to drive ex vivo human stem and progenitors cells into distinct lineages, with the goal being in vivo replacement of particular hematopoietic cell lineages. This approach has recently been used to correct the delayed platelet recovery following UCB stem cell transplantation. Ex vivo culture of CD34$^+$ UCB cells with thrombopoietin (TPO) accelerates platelet production following engraftment into NOD-*scid* mice [130], and this may be a useful approach for improving the production of human platelets in HSC-engrafted NOD-*scid IL2rγ^{null}* mice [48] (Table 4).

Table 4 Examples of ex vivo manipulations of human hematolymphoid cells carried out to improve human engraftment

Treatment	Reference
HSC expansion or upregulation of homing molecules using cytokines	
IL–3, IL–6, SCF, FLT–3 ligand, and IL–11	43
FLT–3 ligand, SCF, megakaryocyte growth development factor (MGDF), and G–CSF	55
G–CSF and MGDF	12
SCF	142
Notch ligand delta1	26
IL–3 and SCF	115
SDF–1 peptide analog, SCF, TPO, Flt–3 ligand, and BM stromal cell culture supernatant	63
Coculture with human brain endothelial cells	24
SCF, GM–CSF, IL–3, IL–6, EPO, and TPO	37
FLT–3 ligand, SCF, and TPO	30
FLT–3 ligand, IL–7, and TPO	61
SCF, FLT–3 ligand, G–CSF, GM–CSF, IL–3, IL–6, MGDF, EPO, and TPO	87
FLT–3 ligand, IL–6, MGDF, and SCF	95
SCF and FLT–3 ligand	141
Non–cytokine–mediated HSC expansion	
Surface fucosylation of HSC	138
Chromatin–modifying agents	4
HSC differentiation	
Myeloid and B lymphoid lineage using hepatocyte growth factor and TPO	40
Megakaryocyte expansion using TPO	130

5 Coinjection of HSC with Mesenchymal Stem Cells and Other Stromal Cell Populations

Engraftment of immunodeficient mice with human HSC provides the potential to generate a full human hematopoietic system, including the immune system. However, as discussed above, numerous human-specific factors required for human hematolymphoid cell engraftment, differentiation, and survival are not produced by human hematopoietic cells. To attempt to overcome this limitation, investigators have coinjected human HSC with human mesenchymal stem cells (MSC), unrestricted somatic stem cells (USSC), or human cytokine-transduced stromal cells in attempts to provide a "stromal" human cell population capable of facilitating human HSC engraftment (Table 5). The hypothesized mechanisms are thought to be establishment of an appropriate stromal microenvironment of human origin, secretion of human-specific factors that are not produced by human hematopoietic cells, or both. Of particular interest has been the relative success of coinjection of allogeneic human MSC plus HSC. Coinjection of these allogeneic cell populations does not appear to lead to a graft-versus-host reaction between these two tissues [94]. MSC are thought to be immunosuppressive [92] and may in fact suppress both the cord blood T cell response to the allogeneic MSC as well as the host innate immune system, effectively facilitating HSC engraftment.

6 The IL2 Receptor Gamma Common Chain Knockout Mouse

Investigators attempting to create humanized mice with functional human immune systems have been re-energized with the recent development of immunodeficient mice bearing targeted mutations at the IL-2 receptor common gamma chain locus ($IL2rg^{null}$). Four independent groups have generated immunodeficient mice bearing $IL2r\gamma^{null}$ mutations on four different strain backgrounds. These are the NOD/Lt-$scid\,IL2rg^{null}$ [48, 106], NOD/Shi-$scid\,IL2rg^{null}$ [50, 139], BALB/c-$Rag2^{tm1Fwa}IL2rg^{null}$ [127], and $H2^d\,Rag2^{tm1Fwa}\,IL2rg^{null}$ strains [38]. In these strains of immunodeficient mice, human HSC have been shown to generate complete human immune systems, including human thymocytes and peripheral mature T and B cells, myeloid cells,

Table 5 Human stromal cell populations coinjected with human hematolymphoid cells

Cell population(s) coinjected with HSC	Reference
Mesenchymal stem cells	3, 47, 81
Adipose stromal cells	54
IL–7–transduced stromal cells	27
Unrestricted somatic stem cells	21

myeloid and plasmacytoid dendritic cells, platelets, and RBC [110]. T cells that develop in these mice have a diverse TCR repertoire and mount antigen-specific IgM and IgG antibody responses after immunization with T-dependent antigens (for recent review see [110]). These studies have used different strains, newborn and adult recipients, and different routes of injection.

In the newborn engraftment model, four different routes of human HSC injection have been reported (Fig. 2). These routes of human HSC injection include intraperitoneal (IP), intravenous (IV) via intracardiac (IC) or facial vein, and intrahepatic (IH). The IP route of human HSC engraftment appears to be suboptimal for achieving high engraftment of human HSC whereas the IC and IH routes appear to be equivalent in engraftment of human HSC (T. Pearson, unpublished observations). Direct comparison of the four different strains of immunodeficient $IL2r\gamma^{null}$ mice and their engraftment by the various routes of HSC injection has not been reported, leaving the optimal strain and route of injection in newborns undetermined.

In the adult engraftment model, two main routes of human HSC injection into immunodeficient mice have been reported. The most commonly used route of injection is the IV route via the tail vein [110]. More recently described routes of injection involve the direct intrafemoral [70] or intratibial [133] injection of human HSC. These latter routes of injection overcome the bone marrow homing requirements of human HSC injected intravenously and have been reported to facilitate human HSC engraftment in NOD-*scid* recipients [70, 133]. The engraftment of human HSC following intrafemoral or intratibial injection into NOD-*scid* $Il2r\gamma^{null}$ mice has not been reported.

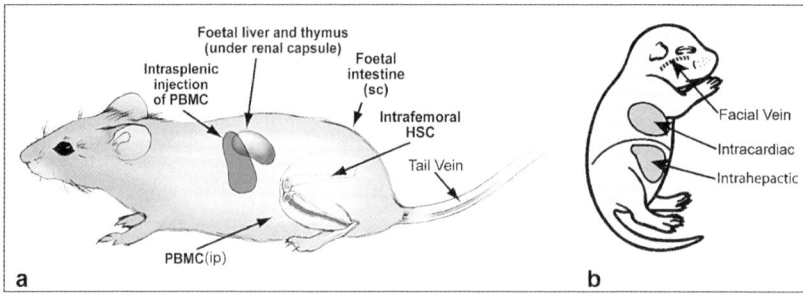

Fig. 2 Major routes of engraftment of human hematopoietic stem cells and tissues in adult and newborn immunodeficient mice. **a** In the adult, numerous routes of stem cell engraftment have been reported depending on the experimental research protocol. Hematopoietic stem cells and been injected using intravenous, intrafemoral, and intratibial injection routes. For fetal tissue engraftment into adults, the traditional site is under the renal capsule. For PBMC engraftment, intraperitoneal, intravenous, and intrasplenic injection routes have been described. **b** For newborn engraftment with HSC, intravenous injection via the facial vein or intracardiac routes have been used, as well as an intrahepatic route of HSC injection

7 Current Research Uses of Humanized Mice

The availability of humanized immunodeficient $Il2r\gamma^{null}$ mice harboring a functional human immune system following engraftment with human HSC has reinvigorated the hope that these humanized mice can be used for the investigation of a number of complex human hematological and immunological processes, including infectious disease, autoimmunity, cancer biology, and regenerative medicine [110]. This aspect of humanized mice has been reviewed extensively since 2005 and is considered only briefly here.

7.1 Infectious Disease

One of the more exciting uses of humanized mice is the investigation of infectious diseases that are human specific [110]. These include viral infections such as human immunodeficiency virus (HIV), Epstein-Barr virus (EBV), hepatitis C, dengue, and the protozoan *Plasmodium falciparum*, the causative agent of malaria and the leading global disease that has killed more individuals than all the wars and other plagues combined (http://en.wikipedia.org/wiki/Plasmodium_falciparum). Because these agents infect human but not mouse cells, no suitable small animal models are currently available for the study of the pathogenesis of these diseases. Humanized mice are also being used to develop immunization protocols that can then be functionally tested by determining their resistance to reinfection with the test organism, a practice that is unethical in humans. Currently, vaccine development requires extensive preclinical testing in nonhuman primates prior to entry into the clinic. Exciting examples of progress in this area are the recent reports on HIV infection in humanized immunodeficient $IL2r\gamma^{null}$ models [6, 10, 134]. These humanized mice have long-lasting infection of human cells by CXCR4- and CCR5-tropic HIV isolates as well as HIV-specific human immune responses.

7.2 Autoimmunity

Animal models have been used extensively to study the pathogenesis of autoimmune diseases. Reliance on animal models is in part recognition of the importance of patient safety-first do no harm. Unlike humans, mice and rats can be experimentally manipulated to study the disease process. Disease can be deliberately induced, the diseased tissue can be biopsied, and the animals can be necropsied at various stages of the autoimmune disease process for study. The genetic basis for the autoimmune disease can be identified, the genome can be altered with knockout, knockin, or transgenic technology, and the genome can be fixed by inbreeding. Importantly, therapies to prevent or reverse the autoimmune disorder can readily be tested without the ethical concerns associated with clinical research in humans.

Humanized mice have been used for the study of autoimmune type 1 diabetes, autoimmune thyroiditis, and rheumatoid arthritis [110]. In diabetes, PBMC from diabetic individuals adoptively transferred to CB17-*scid* mice led to the detection of autoantibodies to islet components, but no infiltration or beta cell destruction was observed [85]. More recently, the development of human T cell clones with specificities to islet autoantigens has permitted the study of the adoptive transfer of diabetes into NOD-*scid* mice. In these studies, infiltration, but not islet cell destruction, was detected [129]. The use of newer models of immunodeficient mice based on the $IL2r\gamma^{null}$ mutation and transgenic expression of human HLA molecules (Table 3) may provide new models that will permit the direct study of human autoreactive T cells in vivo.

7.3 Cancer

Immunodeficient mice have been used to study the growth of tumor cells for over 3 decades [110]. The first studies were the implantation of solid tumors in nude mice [33], followed by the observation that some human lymphomas could grow in CB17-*scid* mice [46]. The development of NOD-*scid* mice permitted the growth of many primary human lymphomas and leukemias that could not grow in CB17-*scid* mice [46], in part because of the decreased innate immunity and lower NK cell activity in NOD-*scid* as compared with CB17-*scid* mice [107]. More recently, primary acute myelogenous leukemia (AML) cells that did not grow in NOD-*scid* mice have been observed to grow in NOD/Lt-*scid* $IL2r\gamma^{null}$ mice [140], suggesting that the in vivo study of primary human leukemias and lymphomas not feasible in previous generations of immunodeficient mice will now be possible.

An important concept that has been validated with immunodeficient mice is the existence of a "tumor stem cell" [83, 93]. This concept has been tested by transplantation of small numbers of tumor cells with stem cell characteristics into immunodeficient mice to document that they self-renew, differentiate, and give rise to the tumor [110]. This concept has translated into important clinical considerations, as the primary target in tumor therapy has previously been reduction of tumor mass, not targeting of tumor stem cells. Furthermore, for tumors such as AML, one treatment option is to isolate CD34$^+$ cells from the patient's bone marrow for autologous transplantation after the disease has been driven into remission. However, the rate of recurrence is high after this procedure, and it has recently been shown that AML tumor stem cells also express CD34 [1, 28, 140], leading to their enrichment, not depletion, in autologous stem cell transplantation protocols,. Indeed, with the use of NOD-*scid* mice as hosts, human tumor stem cells have been isolated for a number of additional malignancies, including myeloma [88] and breast [2], brain [111], and pancreatic [62] cancers. Future studies of human tumor stem cells using immunodeficient $IL2r\gamma^{null}$ mice as the in vivo testing ground for functional analyses could lead to individualized and focused new therapies that specifically target the tumor stem cell on a patient-by-patient basis.

7.4 Regenerative Medicine

Regenerative medicine is a rapidly growing field that is based on using stem cell therapy to replace damaged or destroyed cells and tissue [103, 112]. Important for rapid advancement of this field will be the ability to test the regenerative potential of the stem cells in an in vivo setting before translation of the cell-based therapy to the clinic. To address this need, immunodeficient animal models with genetic or induced tissue injury are being used for in vivo evaluation of the regenerative capacity of human stem cells. The need for tissue damage or injury for stem cells to repair tissue via regeneration, transdifferentiation, or cell fusion is a common feature particularly amenable for modeling in animals as they can easily be experimentally manipulated for study. This approach is being applied for the preclinical testing of human embryonic stem (ES) cell-derived populations, with particular interest focused on the use of ES and HSC [110] as stem cell sources for generating insulin-producing beta cells (http://www.betacell.org/).

Recognition that transdifferentiation and/or cell fusion of transplanted HSC also occurs in damaged tissues has led to intense investigation in this area [90]. Although not a form of "true regeneration of endogenous tissues," clinical efficacy has been observed in animal models in which HSC transplantation leading to cell fusion has been used for the treatment of diseased or damaged tissues [110]. The use of stem cell therapy has also been found to have efficacy in treating heart damage in experimental models [49, 66] and in humans [116]. The use of humanized mouse models to study the regenerative capacity for human stem cells to repair damaged or lost tissues presents the exciting possibility that humanized mice may serve as a "preclinical" bridge for translating data from animal models to human cells and tissues before their application in the clinic (Fig. 3). Important in these experiments will be

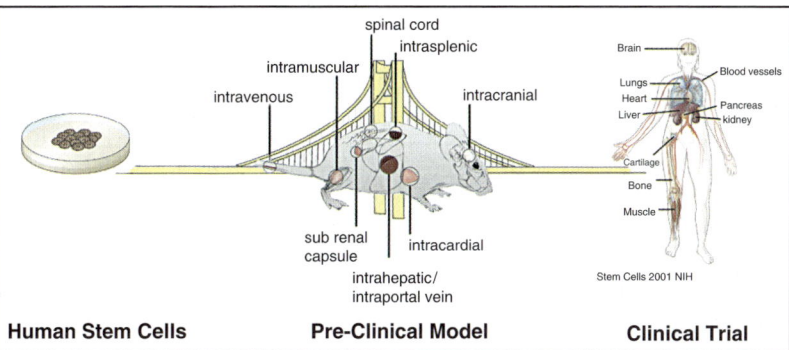

Fig. 3 Humanized SCID mice serve as a preclinical bridge between studies in laboratory animals and clinical research on humans. Stem cells are defined as long-term self-renewing cells that can repopulate the complete hematopoietic system and sustain long-term multilineage hematopoiesis. Functional analyses of these cells can only be accomplished with in vivo model systems. Humanized mouse models are used to define the functional activity of human stem cells and can function as a preclinical bridge between in vitro analyses of stem cells and the potential clinical use of stem cell therapy for a variety of human diseases in the clinic

the use of immunodeficient *IL2rγ*^{null} mice engrafted with human embryonic or adult stem cells. It has recently been reported that ES cells and their progeny, HSC and MSC, are highly susceptible to NK cell-mediated killing [41, 114, 126]. The use of immunodeficient *IL2rγ*^{null} mice that are completely NK cell deficient may permit study of the regenerative potential of stem and progenitor cell populations not possible in previous immunodeficient humanized mouse models.

8 Remaining Hurdles and Future Directions

There are a number of remaining limitations in immunodeficient *IL2rγ*^{null} humanized mouse models, comprising three broad categories. First, there are many human-specific molecules required for proper human immune system function that are not expressed in mouse tissues. Second, there are issues of remaining innate immunity that present obstacles to human stem cell engraftment. Third, the architecture of the lymphoid system remains undeveloped, and lymph nodes and secondary lymphoid organs in unmanipulated immunodeficient *IL2rγ*^{null} mice are exceedingly small and their stromal cell structure is very poorly developed. Overcoming these limitations should be achievable in the future with additional modifications, some of which are outlined below.

8.1 Human-Specific Molecule Expression

Genetic manipulation of mice is a powerful tool that can be used to knock out genes, knock in genes, and transgenically express human genes. In the latter case, expression of human-specific genes required for proper human immune system development and function can easily be accomplished. Of great importance is the expression of human HLA transgenes (Table 3) that will permit proper thymic selection of developing human T cells in the immunodeficient *IL2rγ*^{null} host. Additional molecules, such as cytokines and growth factors including IL-2, IL-7, and BLyS, will facilitate the development, differentiation, and survival of a functional human immune system. IL-2 may be particularly important, as it is a key cytokine in T lymphocyte development and function. Exogenous administration of IL-2 could not be performed in previous generations of immunodeficient mice, particularly those on the NOD-*scid* background, because of IL-2-dependent acceleration of thymic lymphoma development and early mortality [41, 107]. In immunodeficient *IL2rγ*^{null} mice, IL-2 signaling cannot occur and exogenous administration of IL-2 should not induce thymic lymphomas and premature death. Additional molecules, such as adhesion and homing molecules expressed on tissues in the host, may overcome some of the issues associated with homing of human immune cells. Engraftment of human lymphocytes may similarly facilitate restoration of the structure of the secondary lymphoid organs in HSC-engrafted immunodeficient *IL2rγ*^{null} mice.

8.2 New Approaches for Reducing Residual Innate Immunity

Although much improved as recipients of human stem cells and PBMC, immunodeficient *IL2rγ^{null}* mice still have residual innate immune function that impedes engraftment. The residual innate immune compartment, while functionally depressed in the absence of the *IL2rγ* signaling, still includes macrophages, granulocytes, dendritic cells, and Langerhans cells capable of mediating resistance to engraftment with human hematolymphoid cells. Both genetic approaches and exogenous administration of specific reagents can be used to further depress innate immunity to improve human cell engraftment. For example, transgenic simian diphtheria toxic receptor expression on dendritic cells, macrophages, and other cell populations, driven by the *Itgax* (CD11c) [52], *Itgam* (CD11b) [31], (Langerin) [9], or *Cre* [68] promoters, has been generated. Because mouse cells are relatively insensitive to the effects of diphtheria toxin [20], cells transgenically expressing the diphtheria toxic receptor can be selectively depleted in immunodeficient *IL2rγ^{null}* mice by exogenous administration of diphtheria toxin before engraftment of human hematolymphoid cells. Finally, anti-Gr1 mAb may prove useful to deplete host neutrophils before engraftment of human cells [100].

8.3 Artificial Organoids

One of the remaining concerns relating to the humanized immunodeficient *IL2rγ^{null}* mice model is the lymphoid architecture and poor lymph node and secondary peripheral lymphoid organ development. Appropriate lymphoid architecture requires interaction of follicular dendritic cells (FDC) with lymphocytes and other nonlymphoid cells for full development [36]. In immunodeficient *IL2rγ^{null}* mice that lack lymphocytes, it is not surprising that lymphoid architecture is underdeveloped. This may have functional consequences, as secondary lymphoid organs are the primary site for antigen recognition and generation of immune responses. Reconstitution of immunodeficient *IL2rγ^{null}* mice with a human immune system may in part overcome this limitation, as immunization of these mice leading to the population of the lymphoid structures with human lymphocytes may promote lymphoid architecture development and function [134].

Additionally, progress is being made in the development of artificial lymphoid structures. Artificial thymic stroma [89] and biocompatible scaffolding [117] are being developed as approaches for providing the appropriate architectural environment for proper lymphocyte development and function. In addition, transduction of bone marrow or thymic stromal cells with factors that are needed for hematopoiesis or lymphocyte development may provide a source for those factors that are not provided by the murine stromal environment. For example, a recent report suggests that the use of a CXCR4-containing viral vector to drive CXCR4 expression of human HSC enhances homing and engraftment of the stem cells in immunodeficient mice [15].

9 Conclusions

The ability to study complex human biological processes in a small animal model that can be experimentally manipulated offers great promise for rapid advances in many areas of scientific research. The pathway to achieving this goal started approximately 40 years ago, and has progressed in a manner similar to that in other scientific disciplines: a handful of breakthrough advances interspersed in years of hard work. The scientific community has just experienced a significant breakthrough in the field of humanized mice, with the development of immunodeficient $IL2r\gamma^{null}$ mice that support robust human hematolymphoid engraftment. However, limitations remain in the generation of a fully functional human immune system in these mice that recapitulates immune responses observed in immunocompetent individuals. Applying approaches pioneered to improve HSC engraftment and function in earlier generations of immunodeficient hosts (i.e., the steady hard work done between the big leaps in progress) may prove to be a fertile field for clearing the remaining hurdles in today's immunodeficient murine hosts. As these remaining limitations are overcome and improvements in the models are achieved, the use of humanized mice as robust preclinical models for bridging the route from bench to bedside will become increasingly important (Fig. 3).

Acknowledgements We thank Drs. Thomas Chase, Bonnie Lyons, and Aldo Rossini for critical review of this chapter. Supported by the Beta Cell Biology Consortium and Autoimmunity Prevention Centers from NIH, the Juvenile Diabetes Research Foundation, International, the American Diabetes Association, the Diabetes Endocrinology Center, the National Cancer Institute, the National Institute of Allergy and Infectious Diseases, and the National Heart, Lung, and Blood Institute of NIH. The contents of this publication are solely the responsibility of the authors and do not necessarily represent the official view of the National Institutes of Health.

References

1. Ailles LE, Gerhard B, Kawagoe H and Hogge DE (1999) Growth characteristics of acute myelogenous leukemia progenitors that initiate malignant hematopoiesis in nonobese diabetic/severe combined immunodeficient mice. Blood 94:1761-72
2. Al-Hajj M, Wicha MS, Benito-Hernandez A, Morrison SJ and Clarke MF (2003) Prospective identification of tumorigenic breast cancer cells. Proc Natl Acad Sci USA 100:3983-8
3. Angelopoulou M, Novelli E, Grove JE, Rinder HM, Civin C, Cheng L and Krause DS (2003) Cotransplantation of human mesenchymal stem cells enhances human myelopoiesis and megakaryocytopoiesis in NOD/SCID mice. Exp Hematol 31:413-20
4. Araki H, Mahmud N, Milhem M, Nunez R, Xu M, Beam CA and Hoffman R (2006) Expansion of human umbilical cord blood SCID-repopulating cells using chromatin-modifying agents. Exp Hematol 34:140-9
5. Auffray I, Dubart A, Izac B, Vainchenker W and Coulombel L (1994) A murine stromal cell line promotes the proliferation of the human factor-dependent leukemic cell line UT-7. Exp Hematol 22:417-24
6. Baenziger S, Tussiwand R, Schlaepfer E, Mazzucchelli L, Heikenwalder M, Kurrer MO, Behnke S, Frey J, Oxenius A, Joller H, Aguzzi A, Manz MG and Speck RF (2006) Disseminated

and sustained HIV infection in CD34+ cord blood cell-transplanted Rag2–/–gamma c–/– mice. Proc Natl Acad Sci USA 103:15951-6
7. Baiocchi RA, Ward JS, Carrodeguas L, Eisenbeis CF, Peng R, Roychowdhury S, Vourganti S, Sekula T, O'Brien M, Moeschberger M and Caligiuri MA (2001) GM-CSF and IL-2 induce specific cellular immunity and provide protection against Epstein-Barr virus lymphoproliferative disorder. J Clin Invest 108:887-94
8. Banuelos SJ, Shultz LD, Greiner DL, Burzenski LM, Gott B, Lyons BL, Rossini AA and Appel MC (2004) Rejection of human islets and human HLA-A2.1 transgenic mouse islets by alloreactive human lymphocytes in immunodeficient NOD-scid and NOD-Rag1nullPrf1null mice. Clin Immunol 112:273-283
9. Bennett CL, van Rijn E, Jung S, Inaba K, Steinman RM, Kapsenberg ML and Clausen BE (2005) Inducible ablation of mouse Langerhans cells diminishes but fails to abrogate contact hypersensitivity. J Cell Biol 169:569-76
10. Berges BK, Wheat WH, Palmer BE, Connick E and Akkina R (2006) HIV-1 infection and CD4 T cell depletion in the humanized Rag2–/–gamma c–/– (RAG-hu) mouse model. Retrovirology 3:76
11. Bock TA, Orlic D, Dunbar CE, Broxmeyer HE and Bodine DM (1995) Improved engraftment of human hematopoietic cells in severe combined immunodeficient (SCID) mice carrying human cytokine transgenes. J Exp Med 182:2037-43
12. Boiron JM, Dazey B, Cailliot C, Launay B, Attal M, Mazurier F, McNiece IK, Ivanovic Z, Caraux J, Marit G and Reiffers J (2006) Large-scale expansion and transplantation of CD34$^+$ hematopoietic cells: in vitro and in vivo confirmation of neutropenia abrogation related to the expansion process without impairment of the long-term engraftment capacity. Transfusion 46:1934-42
13. Bornstein R, Flores AI, Montalban MA, del Rey MJ, de la Serna J and Gilsanz F (2005) A modified cord blood collection method achieves sufficient cell levels for transplantation in most adult patients. Stem Cells 23:324-34
14. Bosma GC, Custer RP and Bosma MJ (1983) A severe combined immunodeficiency mutation in the mouse. Nature 301:527-30
15. Brenner S, Whiting-Theobald N, Kawai T, Linton GF, Rudikoff AG, Choi U, Ryser MF, Murphy PM, Sechler JM and Malech HL (2004) CXCR4-transgene expression significantly improves marrow engraftment of cultured hematopoietic stem cells. Stem Cells 22:1128-33
16. Camacho RE, Wnek R, Shah K, Zaller DM, O'Reilly RJ, Collins N, Fitzgerald-Bocarsly P and Koo GC (2004) Intra-thymic/splenic engraftment of human T cells in HLA-DR1 transgenic NOD/scid mice. Cell Immunol 232:86-95
17. Cao T and Leroux-Roels G (2000) Antigen-specific T cell responses in human peripheral blood leucocyte (hu-PBL)-mouse chimera conditioned with radiation and an antibody directed against the mouse IL-2 receptor beta-chain. Clin Exp Immunol 122:117-23
18. Carballido JM, Schols D, Namikawa R, Zurawski S, Zurawski G, Roncarolo MG and de Vries JE (1995) IL-4 induces human B cell maturation and IgE synthesis in SCID-hu mice. Inhibition of ongoing IgE production by in vivo treatment with an IL-4/IL-13 receptor antagonist. J Immunol 155:4162-70
19. Cashman JD and Eaves CJ (1999) Human growth factor-enhanced regeneration of transplantable human hematopoietic stem cells in nonobese diabetic/severe combined immunodeficient mice. Blood 93:481-7
20. Cha JH, Chang MY, Richardson JA and Eidels L (2003) Transgenic mice expressing the diphtheria toxin receptor are sensitive to the toxin. Mol Microbiol 49:235-40
21. Chan SL, Choi M, Wnendt S, Kraus M, Teng E, Leong HF and Merchav S (2007) Enhanced in vivo homing of uncultured and selectively amplified cord blood CD34+ cells by cotransplantation with cord blood-derived unrestricted somatic stem cells. Stem Cells 25:529-36
22. Christianson SW, Greiner DL, Schweitzer IB, Gott B, Beamer GL, Schweitzer PA, Hesselton RA and Shultz LD (1996) Role of natural killer cells on engraftment of human lymphoid cells and on metastasis of human T-lymphoblastoid leukemia cells in C57BL/6J-*scid* mice and in C57BL/6J-*scid bg* mice. Cell Immunol 171:186-199

23. Christianson SW, Greiner DL, Hesselton RA, Leif JH, Wagar EJ, Schweitzer IB, Rajan TV, Gott B, Roopenian DC and Shultz LD (1997) Enhanced human CD4+ T cell engraftment in beta2-microglobulin-deficient NOD-scid mice. J Immunol 158:3578-86
24. Chute JP, Saini AA, Chute DJ, Wells MR, Clark WB, Harlan DM, Park J, Stull MK, Civin C and Davis TA (2002) Ex vivo culture with human brain endothelial cells increases the SCID-repopulating capacity of adult human bone marrow. Blood 100:4433-9
25. Dao MA, Shah AJ, Crooks GM and Nolta JA (1998) Engraftment and retroviral marking of CD34+ and CD34+CD38– human hematopoietic progenitors assessed in immune-deficient mice. Blood 91:1243-55
26. Delaney C, Varnum-Finney B, Aoyama K, Brashem-Stein C and Bernstein ID (2005) Dose-dependent effects of the Notch ligand Delta1 on ex vivo differentiation and in vivo marrow repopulating ability of cord blood cells. Blood 106:2693-9
27. Di Ianni M, Papa BD, De Ioanni M, Terenzi A, Sportoletti P, Moretti L, Falzetti F, Gaozza E, Zei T, Spinozzi F, Bagnis C, Mannoni P, Bonifacio E, Falini B, Martelli MF and Tabilio A (2005) Interleukin 7-engineered stromal cells: a new approach for hastening naive T cell recruitment. Hum Gene Ther 16:752-64
28. Dick JE and Lapidot T (2005) Biology of normal and acute myeloid leukemia stem cells. Int J Hematol 82:389-96
29. DiSanto JP, Muller W, Guy-Grand D, Fischer A and Rajewsky K (1995) Lymphoid development in mice with a targeted deletion of the interleukin 2 receptor gamma chain. Proc Natl Acad Sci USA 92:377-81
30. Dravid G and Rao SG (2002) Ex vivo expansion of stem cells from umbilical cord blood: expression of cell adhesion molecules. Stem Cells 20:183-9
31. Duffield JS, Forbes SJ, Constandinou CM, Clay S, Partolina M, Vuthoori S, Wu S, Lang R and Iredale JP (2005) Selective depletion of macrophages reveals distinct, opposing roles during liver injury and repair. J Clin Invest 115:56-65
32. Faulkner L, Borysiewicz LK and Man S (1998) The use of human leucocyte antigen class I transgenic mice to investigate human immune function. J Immunol Methods 221:1-16
33. Fogh J, Fogh JM and Orfeo T (1977) One hundred and twenty-seven cultured human tumor cell lines producing tumors in nude mice. J Natl Cancer Inst 59:221-6
34. Fraser CC, Chen BP, Webb S, van Rooijen N and Kraal G (1995) Circulation of human hematopoietic cells in severe combined immunodeficient mice after Cl2MDP-liposome-mediated macrophage depletion. Blood 86:183-92
35. Friese MA, Jensen LT, Willcox N and Fugger L (2006) Humanized mouse models for organ-specific autoimmune diseases. Curr Opin Immunol 18:704-9
36. Fu YX, Huang G, Matsumoto M, Molina H and Chaplin DD (1997) Independent signals regulate development of primary and secondary follicle structure in spleen and mesenteric lymph node. Proc Natl Acad Sci USA 94:5739-43
37. Gammaitoni L, Bruno S, Sanavio F, Gunetti M, Kollet O, Cavalloni G, Falda M, Fagioli F, Lapidot T, Aglietta M and Piacibello W (2003) Ex vivo expansion of human adult stem cells capable of primary and secondary hemopoietic reconstitution. Exp Hematol 31:261-70
38. Gimeno R, Weijer K, Voordouw A, Uittenbogaart CH, Legrand N, Alves NL, Wijnands E, Blom B and Spits H (2004) Monitoring the effect of gene silencing by RNA interference in human CD34+ cells injected into newborn RAG2−/− gammac−/− mice: functional inactivation of p53 in developing T cells. Blood 104:3886-93
39. Gotoh M, Takasu H, Harada K and Yamaoka T (2002) Development of HLA-A2402/K(b) transgenic mice. Int J Cancer 100:565-70
40. Grassinger J, Mueller G, Zaiss M, Kunz-Schughart LA, Andreesen R and Hennemann B (2006) Differentiation of hematopoietic progenitor cells towards the myeloid and B-lymphoid lineage by hepatocyte growth factor (HGF) and thrombopoietin (TPO) together with early acting cytokines. Eur J Haematol 77:134-44
41. Greiner DL, Hesselton RA and Shultz LD (1998) SCID mouse models of human stem cell engraftment. Stem Cells 16:166-77

42. Greiner DL and Shultz LD (1998) The Use of NOD/LtSz-scid/scid mice in biomedical research. In: Leiter E and Atkinson M (eds) NOD Mice and Related Strains: Research Applications in Diabetes, AIDS, Cancer and Other Diseases. Landes Bioscience, Austin, TX
43. Guenechea G, Segovia JC, Albella B, Lamana M, Ramirez M, Regidor C, Fernandez MN and Bueren JA (1999) Delayed engraftment of nonobese diabetic/severe combined immunodeficient mice transplanted with ex vivo-expanded human CD34+ cord blood cells. Blood 93:1097-105
44. Hesselton RM, Greiner DL, Mordes JP, Rajan TV, Sullivan JL and Shultz LD (1995) High levels of human peripheral blood mononuclear cell engraftment and enhanced susceptibility to HIV-1 infection in NOD/LtSz-*scid/scid* mice. J Inf Dis 172:774-782
45. Hidalgo A and Frenette PS (2005) Enforced fucosylation of neonatal CD34+ cells generates selectin ligands that enhance the initial interactions with microvessels but not homing to bone marrow. Blood 105:567-75
46. Hudson WA, Li Q, Le C and Kersey JH (1998) Xenotransplantation of human lymphoid malignancies is optimized in mice with multiple immunologic defects. Leukemia 12:2029-33
47. in 't Anker PS, Noort WA, Kruisselbrink AB, Scherjon SA, Beekhuizen W, Willemze R, Kanhai HH and Fibbe WE (2003) Nonexpanded primary lung and bone marrow-derived mesenchymal cells promote the engraftment of umbilical cord blood-derived CD34+ cells in NOD/SCID mice. Exp Hematol 31:881-9
48. Ishikawa F, Yasukawa M, Lyons B, Yoshida S, Miyamoto T, Yoshimoto G, Watanabe T, Akashi K, Shultz LD and Harada M (2005) Development of functional human blood and immune systems in NOD/SCID/IL2 receptor γ chain^{null} mice. Blood 106:1565-73
49. Ishikawa F, Shimazu H, Shultz LD, Fukata M, Nakamura R, Lyons B, Shimoda K, Shimoda S, Kanemaru T, Nakamura K, Ito H, Kaji Y, Perry AC and Harada M (2006) Purified human hematopoietic stem cells contribute to the generation of cardiomyocytes through cell fusion. FASEB J 20:950-2
50. Ito M, Hiramatsu H, Kobayashi K, Suzue K, Kawahata M, Hioki K, Ueyama Y, Koyanagi Y, Sugamura K, Tsuji K, Heike T and Nakahata T (2002) NOD/SCID/γ_c^{null} mouse: an excellent recipient mouse model for engraftment of human cells. Blood 100:3175-82
51. Jacobs H, Krimpenfort P, Haks M, Allen J, Blom B, Demolliere C, Kruisbeek A, Spits H and Berns A (1999) PIM1 reconstitutes thymus cellularity in interleukin 7- and common gamma chain-mutant mice and permits thymocyte maturation in Rag- but not CD3gamma-deficient mice. J Exp Med 190:1059-68
52. Jung S, Unutmaz D, Wong P, Sano G, De los Santos K, Sparwasser T, Wu S, Vuthoori S, Ko K, Zavala F, Pamer EG, Littman DR and Lang RA (2002) In vivo depletion of CD11c+ dendritic cells abrogates priming of CD8+ T cells by exogenous cell-associated antigens. Immunity 17:211-20
53. Kapp U, Bhatia M, Bonnet D, Murdoch B and Dick JE (1998) Treatment of non-obese diabetic (NOD)/Severe-combined immunodeficient mice (SCID) with flt3 ligand and interleukin-7 impairs the B-lineage commitment of repopulating cells after transplantation of human hematopoietic cells. Blood 92:2024-31
54. Kim SJ, Cho HH, Kim YJ, Seo SY, Kim HN, Lee JB, Kim JH, Chung JS and Jung JS (2005) Human adipose stromal cells expanded in human serum promote engraftment of human peripheral blood hematopoietic stem cells in NOD/SCID mice. Biochem Biophys Res Commun 329:25-31
55. Kobari L, Pflumio F, Giarratana M, Li X, Titeux M, Izac B, Leteurtre F, Coulombel L and Douay L (2000) In vitro and in vivo evidence for the long-term multilineage (myeloid, B, NK, and T) reconstitution capacity of ex vivo expanded human CD34+ cord blood cells. Exp Hematol 28:1470-80
56. Kollet O, Aviram R, Chebath J, ben-Hur H, Nagler A, Shultz L, Revel M and Lapidot T (1999) The soluble interleukin-6 (IL-6) receptor/IL-6 fusion protein enhances in vitro maintenance

and proliferation of human $CD34^+CD38^{-/low}$ cells capable of repopulating severe combined immunodeficiency mice. Blood 94:923-31
57. Krimpenfort P, Rudenko G, Hochstenbach F, Guessow D, Berns A and Ploegh H (1987) Crosses of two independently derived transgenic mice demonstrate functional complementation of the genes encoding heavy (HLA-B27) and light (beta2-microglobulin) chains of HLA class I antigens. EMBO J 6:1673-6
58. Lapenta C, Santini SM, Spada M, Donati S, Urbani F, Accapezzato D, Franceschini D, Andreotti M, Barnaba V and Belardelli F (2006) IFN-alpha-conditioned dendritic cells are highly efficient in inducing cross-priming $CD8^+$ T cells against exogenous viral antigens. Eur J Immunol 36:2046-60
59. Lapidot T, Pflumio F, Deodens M, Murdoch B, Williams DE and Dick JE (1992) Cytokine stimulation of multi lineage hematopoiesis from immature human cells engrafted in SCID mice. Science 255:1137-1141
60. Leonard WJ (1996) Dysfunctional cytokine receptor signaling in severe combined immunodeficiency. J Investig Med 44:304-11
61. Lewis ID, Almeida-Porada G, Du J, Lemischka IR, Moore KA, Zanjani ED and Verfaillie CM (2001) Umbilical cord blood cells capable of engrafting in primary, secondary, and tertiary xenogeneic hosts are preserved after ex vivo culture in a noncontact system. Blood 97:3441-9
62. Li C, Heidt DG, Dalerba P, Burant CF, Zhang L, Adsay V, Wicha M, Clarke MF and Simeone DM (2007) Identification of pancreatic cancer stem cells. Cancer Res 67:1030-7
63. Li K, Chuen CK, Lee SM, Law P, Fok TF, Ng PC, Li CK, Wong D, Merzouk A, Salari H, Gu GJ and Yuen PM (2006) Small peptide analogue of SDF-1alpha supports survival of cord blood CD34+ cells in synergy with other cytokines and enhances their ex vivo expansion and engraftment into nonobese diabetic/severe combined immunodeficient mice. Stem Cells 24:55-64
64. Liu J, Purdy LE, Rabinovitch S, Jevnikar AM and Elliott JF (1999) Major DQ8-restricted T-cell epitopes for human GAD65 mapped using human CD4, DQA1*0301, DQB1*0302 transgenic IA^{null} NOD mice. Diabetes 48:469-77
65. Lowry PA, Shultz LD, Greiner DL, Hesselton RM, Kittler EL, Tiarks CY, Rao SS, Reilly J, Leif JH, Ramshaw H, Stewart FM and Quesenberry PJ (1996) Improved engraftment of human cord blood stem cells in NOD/LtSz- scid/scid mice after irradiation or multiple-day injections into unirradiated recipients. Biol Blood Marrow Transplant 2:15-23
66. Ma N, Ladilov Y, Kaminski A, Piechaczek C, Choi YH, Li W, Steinhoff G and Stamm C (2006) Umbilical cord blood cell transplantation for myocardial regeneration. Transplant Proc 38:771-3
67. Mangalam A, Rodriguez M and David C (2006) Role of MHC class II expressing CD4+ T cells in proteolipid protein91-110-induced EAE in HLA-DR3 transgenic mice. Eur J Immunol 36:3356-70
68. Matsumura H, Hasuwa H, Inoue N, Ikawa M and Okabe M (2004) Lineage-specific cell disruption in living mice by Cre-mediated expression of diphtheria toxin A chain. Biochem Biophys Res Commun 321:275-9
69. May KF, Jr., Roychowdhury S, Bhatt D, Kocak E, Bai XF, Liu JQ, Ferketich AK, Martin EW, Jr., Caligiuri MA, Zheng P and Liu Y (2005) Anti-human CTLA-4 monoclonal antibody promotes T-cell expansion and immunity in a hu-PBL-SCID model: a new method for preclinical screening of costimulatory monoclonal antibodies. Blood 105:1114-20
70. Mazurier F, Doedens M, Gan OI and Dick JE (2003) Rapid myeloerythroid repopulation after intrafemoral transplantation of NOD-SCID mice reveals a new class of human stem cells. Nat Med 9:959-63
71. McCune JM, Namikawa R, Kaneshima H, Shultz LD, Lieberman M and Weissman IL (1988) The SCID-hu mouse: murine model for the analysis of human hematolymphoid differentiation and function. Science 241:1632-9
72. McKenzie JL, Gan OI, Doedens M and Dick JE (2005) Human short-term repopulating stem cells are efficiently detected following intrafemoral transplantation into NOD/SCID recipients depleted of CD122+ cells. Blood 106:1259-61

73. Mercurio AM, Schwarting GA and Robbins PW (1984) Glycolipids of the mouse peritoneal macrophage. J. Exp. Med. 160:1114-1125
74. Mosier DE, Gulizia RJ, Baird SM and Wilson DB (1988) Transfer of a functional human immune system to mice with severe combined immunodeficiency. Nature 335:256-9
75. Murphy WJ, Durum SK and Longo DL (1992) Human growth hormone promotes engraftment of murine or human T cells in severe combined immunodeficient mice. Proc Natl Acad Sci USA 89:4481-5
76. Murphy WJ, Funakoshi S, Fanslow WC, Rager HC, Taub DD and Longo DL (1999) CD40 stimulation promotes human secondary immunoglobulin responses in HuPBL-SCID chimeras. Clin Immunol 90:22-7
77. Nabozny GH, Baisch JM, Cheng S, Cosgrove D, Griffiths MM, Luthra HS and David CS (1996) HLA-DQ8 transgenic mice are highly susceptible to collagen-induced arthritis: a novel model for human polyarthritis. J Exp Med 183:27-37
78. Nakamura T, Miyakawa Y, Miyamura A, Yamane A, Suzuki H, Ito M, Ohnishi Y, Ishiwata N, Ikeda Y and Tsuruzoe N (2006) A novel nonpeptidyl human c-Mpl activator stimulates human megakaryopoiesis and thrombopoiesis. Blood 107:4300-7
79. Napolitano LA, Stoddart CA, Hanley MB, Wieder E and McCune JM (2003) Effects of IL-7 on early human thymocyte progenitor cells in vitro and in SCID-hu Thy/Liv mice. J Immunol 171:645-54
80. Nicolini FE, Cashman JD, Hogge DE, Humphries RK and Eaves CJ (2004) NOD/SCID mice engineered to express human IL-3, GM-CSF and Steel factor constitutively mobilize engrafted human progenitors and compromise human stem cell regeneration. Leukemia 18:341-7
81. Noort WA, Kruisselbrink AB, in't Anker PS, Kruger M, van Bezooijen RL, de Paus RA, Heemskerk MH, Lowik CW, Falkenburg JH, Willemze R and Fibbe WE (2002) Mesenchymal stem cells promote engraftment of human umbilical cord blood-derived CD34+ cells in NOD/SCID mice. Exp Hematol 30:870-8
82. Ohbo K, Suda T, Hashiyama M, Mantani A, Ikebe M, Miyakawa K, Moriyama M, Nakamura M, Katsuki M, Takahashi K, Yamamura K and Sugamura K (1996) Modulation of hematopoiesis in mice with a truncated mutant of the interleukin-2 receptor gamma chain. Blood 87:956-67
83. Pardal R, Clarke MF and Morrison SJ (2003) Applying the principles of stem-cell biology to cancer. Nat Rev Cancer 3:895-902
84. Perez LE, Alpdogan O, Shieh JH, Wong D, Merzouk A, Salari H, O'Reilly RJ, van den Brink MR and Moore MA (2004) Increased plasma levels of stromal-derived factor-1 (SDF-1/CXCL12) enhance human thrombopoiesis and mobilize human colony-forming cells (CFC) in NOD/SCID mice. Exp Hematol 32:300-7
85. Petersen JS, Marshall MO, Baekkeskov S, Hejnaes KR, Hoier-Madsen M and Dyrberg T (1993) Transfer of type 1 (insulin-dependent) diabetes mellitus associated autoimmunity to mice with severe combined immunodeficiency (SCID). Diabetologia 36:510-5
86. Pflumio F, Izac B, Katz A, Shultz LD, Vainchenker W and Coulombel L (1996) Phenotype and function of human hematopoietic cells engrafting immune-deficient CB17-severe combined immunodeficiency mice and nonobese diabetic-severe combined immunodeficiency mice after transplantation of human cord blood mononuclear cells. Blood 88:3731-40
87. Piacibello W, Sanavio F, Severino A, Dane A, Gammaitoni L, Fagioli F, Perissinotto E, Cavalloni G, Kollet O, Lapidot T and Aglietta M (1999) Engraftment in nonobese diabetic severe combined immunodeficient mice of human CD34+ cord blood cells after ex vivo expansion: evidence for the amplification and self-renewal of repopulating stem cells. Blood 93:3736-49
88. Pilarski LM and Belch AR (2002) Clonotypic myeloma cells able to xenograft myeloma to nonobese diabetic severe combined immunodeficient mice copurify with CD34+ hematopoietic progenitors. Clin Cancer Res 8:3198-204
89. Poznansky MC, Evans RH, Foxall RB, Olszak IT, Piascik AH, Hartman KE, Brander C, Meyer TH, Pykett MJ, Chabner KT, Kalams SA, Rosenzweig M and Scadden DT (2000)

Efficient generation of human T cells from a tissue-engineered thymic organoid. Nat Biotechnol 18:729-34
90. Quaini F, Urbanek K, Beltrami AP, Finato N, Beltrami CA, Nadal-Ginard B, Kajstura J, Leri A and Anversa P (2002) Chimerism of the transplanted heart. N Engl J Med 346:5-15
91. Raju R, Munn SR, Majoribanks C and David CS (1998) Islet cell autoimmunity in NOD mice transgenic for HLA-DQ8 and lacking I-Ag7. Transplant Proc 30:561
92. Ramasamy R, Fazekasova H, Lam EW, Soeiro I, Lombardi G and Dazzi F (2007) Mesenchymal stem cells inhibit dendritic cell differentiation and function by preventing entry into the cell cycle. Transplantation 83:71-6
93. Reya T, Morrison SJ, Clarke MF and Weissman IL (2001) Stem cells, cancer, and cancer stem cells. Nature 414:105-11
94. Ringden O, Uzunel M, Rasmusson I, Remberger M, Sundberg B, Lonnies H, Marschall HU, Dlugosz A, Szakos A, Hassan Z, Omazic B, Aschan J, Barkholt L and Le Blanc K (2006) Mesenchymal stem cells for treatment of therapy-resistant graft-versus-host disease. Transplantation 81:1390-7
95. Rollini P, Kaiser S, Faes-van't Hull E, Kapp U and Leyvraz S (2004) Long-term expansion of transplantable human fetal liver hematopoietic stem cells. Blood 103:1166-70
96. Roychowdhury S, Blaser BW, Freud AG, Katz K, Bhatt D, Ferketich AK, Bergdall V, Kusewitt D, Baiocchi RA and Caligiuri MA (2005) IL-15 but not IL-2 rapidly induces lethal xenogeneic graft-versus-host disease. Blood 106:2433-5
97. Rozemuller H, Knaan-Shanzer S, Hagenbeek A, van Bloois L, Storm G and Martens AC (2004) Enhanced engraftment of human cells in RAG2/gammac double-knockout mice after treatment with CL2MDP liposomes. Exp Hematol 32:1118-25
98. Samira S, Ferrand C, Peled A, Nagler A, Tovbin Y, Ben-Hur H, Taylor N, Globerson A and Lapidot T (2004) Tumor necrosis factor promotes human T-cell development in nonobese diabetic/severe combined immunodeficient mice. Stem Cells 22:1085-100
99. Sandhu J, Shpitz B, Gallinger S and Hozumi N (1994) Human primary immune response in SCID mice engrafted with human peripheral blood lymphocytes. J Immunol 152:3806-13
100. Santini SM, Spada M, Parlato S, Logozzi M, Lapenta C, Proietti E, Belardelli F and Fais S (1998) Treatment of severe combined immunodeficiency mice with anti-murine granulocyte monoclonal antibody improves human leukocyte xenotransplantation. Transplantation 65:416-20
101. Santini SM, Lapenta C, Logozzi M, Parlato S, Spada M, Di Pucchio T and Belardelli F (2000) Type I interferon as a powerful adjuvant for monocyte-derived dendritic cell development and activity in vitro and in Hu-PBL-SCID mice. J Exp Med 191:1777-88
102. Senpuku H, Asano T, Matin K, Salam MA, Tsuha Y, Horibata S, Shimazu Y, Soeno Y, Aoba T, Sata T, Hanada N and Honda M (2002) Effects of human interleukin-18 and interleukin-12 treatment on human lymphocyte engraftment in NOD-scid mouse. Immunology 107:232-42
103. Serakinci N and Keith WN (2006) Therapeutic potential of adult stem cells. Eur J Cancer 42:1243-6
104. Shibata S, Asano T, Noguchi A, Naito M, Ogura A and Doi K (1998) Peritoneal macrophages play an important role in eliminating human cells from severe combined immunodeficient mice transplanted with human peripheral blood lymphocytes. Immunology 93:524-32
105. Shpitz B, Chambers CA, Singhal AB, Hozumi N, Fernandes BJ, Roifman CM, Weiner LM, Roder JC and Gallinger S (1994) High level of function engraftment of severe combined immunodeficient mice with human peripheral blood lymphocytes following pretreatment with radiation and anti-asialo-GM-1. J Immunol Methods 169:1-15
106. Shultz L, Lyons B, Burzenski L, Gott B, Chen X, Chaleff S, Kotb M, Gillies S, King M, J M, Greiner D and Handgretinger R (2005) Human lymphoid and myeloid cell development in NOD/LtSz-*scid IL2rg null* mice engrafted with mobilized human hematopoietic stem cells. J Immunol 174:6477-6489

107. Shultz LD, Schweitzer PA, Christianson SW, Gott B, Schweitzer IB, Tennent B, McKenna S, Mobraaten L, Rajan TV, Greiner DL et al. (1995) Multiple defects in innate and adaptive immunologic function in NOD/LtSz-scid mice. J Immunol 154:180-91
108. Shultz LD, Lang PA, Christianson SW, Gott B, Lyons B, Umeda S, Leiter E, Hesselton R, Wagar EJ, Leif JH, Kollet O, Lapidot T and Greiner DL (2000) NOD/LtSz-Rag1null mice: an immunodeficient and radioresistant model for engraftment of human hematolymphoid cells, HIV infection, and adoptive transfer of NOD mouse diabetogenic T cells. J Immunol 164:2496-507
109. Shultz LD, Banuelos S, Lyons B, Samuels R, Burzenski L, B. G, Land P, Leif J, M. A, A. R and Greiner DL (2003) NOD/LtSz-*Rag1nullPfpnull* mice: a new model system to increase levels of human peripheral leukocyte and hematopoietic stem cell engraftment. Transplantation 76:1036-1042
110. Shultz LD, Ishikawa F and Greiner DL (2007) Humanized mice in translational biomedical research. Nat Rev Immunol 7:118-30
111. Singh SK, Hawkins C, Clarke ID, Squire JA, Bayani J, Hide T, Henkelman RM, Cusimano MD and Dirks PB (2004) Identification of human brain tumour initiating cells. Nature 432:396-401
112. Solter D (2006) From teratocarcinomas to embryonic stem cells and beyond: a history of embryonic stem cell research. Nat Rev Genet 7:319-27
113. Sorrentino BP (2004) Clinical strategies for expansion of haematopoietic stem cells. Nat Rev Immunol 4:878-88
114. Sotiropoulou PA, Perez SA, Gritzapis AD, Baxevanis CN and Papamichail M (2006) Interactions between human mesenchymal stem cells and natural killer cells. Stem Cells 24:74-85
115. Spence SE, Keller JR, Ruscetti FW, McCauslin CS, Gooya JM, Funakoshi S, Longo DL and Murphy WJ (1998) Engraftment of ex vivo expanded and cycling human cord blood hematopoietic progenitor cells in SCID mice. Exp Hematol 26:507-14
116. Stamm C, Liebold A, Steinhoff G and Strunk D (2006) Stem cell therapy for ischemic heart disease: beginning or end of the road? Cell Transplant 15 Suppl 1:S47-S56
117. Suematsu S and Watanabe T (2004) Generation of a synthetic lymphoid tissue-like organoid in mice. Nat Biotechnol 22:1539-45
118. Sugamura K, Asao H, Kondo M, Tanaka N, Ishii N, Ohbo K, Nakamura M and Takeshita T (1996) The interleukin-2 receptor gamma chain: its role in the multiple cytokine receptor complexes and T cell development in XSCID. Annu Rev Immunol 14:179-205
119. Suttles JG, Schwarting GA and Stout RD (1986) Flow cytometric analysis reveals the presence of asialo-Gm on the surface membrane of alloimmune cytotoxic T lymphocytes. J Immunol 136:1586-1591
120. Takaki T, Marron MP, Mathews CE, Guttmann ST, Bottino R, Trucco M, DiLorenzo TP and Serreze DV (2006) HLA-A*0201-restricted T cells from humanized NOD mice recognize autoantigens of potential clinical relevance to type 1 diabetes. J Immunol 176:3257-65
121. Tanaka T, Tsudo M, Karasuyama H, Kitamura F, Kono T, Hatakeyama M, Taniguchi T and Miyasaka M (1991) A novel monoclonal antibody against murine IL-2 receptor beta-chain. Characterization of receptor expression in normal lymphoid cells and EL-4 cells. J Immunol 147:2222-8
122. Taneja V, Behrens M, Mangalam A, Griffiths MM, Luthra HS and David CS (2007) New humanized HLA-DR4-transgenic mice that mimic the sex bias of rheumatoid arthritis. Arthritis Rheum 56:69-78
123. Taub DD, Tsarfaty G, Lloyd AR, Durum SK, Longo DL and Murphy WJ (1994) Growth hormone promotes human T cell adhesion and migration to both human and murine matrix proteins in vitro and directly promotes xenogeneic engraftment. J Clin Invest 94:293-300
124. Taurog JD, Hammer RE, Maika SD, Sams K, El-zaatra FAK, Stimpson SA and Schwab JF (1990) HLA-A27 transgenic mice as potential models of human diseases. In: David CS (eds) Transgenic Mice and Mutants in MHC Research. Springer-Verlag, New York, pp 268-275

125. Terpstra W, Leenen PJ, van den Bos C, Prins A, Loenen WA, Verstegen MM, van Wyngaardt S, van Rooijen N, Wognum AW, Wagemaker G, Wielenga JJ and Lowenberg B (1997) Facilitated engraftment of human hematopoietic cells in severe combined immunodeficient mice following a single injection of Cl2MDP liposomes. Leukemia 11:1049-54
126. Tian X, Woll PS, Morris JK, Linehan JL and Kaufman DS (2006) Hematopoietic engraftment of human embryonic stem cell-derived cells is regulated by recipient innate immunity. Stem Cells 24:1370-80
127. Traggiai E, Chicha L, Mazzucchelli L, Bronz L, Piffaretti JC, Lanzavecchia A and Manz MG (2004) Development of a human adaptive immune system in cord blood cell-transplanted mice. Science 304:104-7
128. Uze G, Lutfalla G and Gresser I (1990) Genetic transfer of a functional human interferon alpha receptor into mouse cells: cloning and expression of its cDNA. Cell 60:225-34
129. van Halteren AG, Kardol MJ, Mulder A and Roep BO (2005) Homing of human autoreactive T cells into pancreatic tissue of NOD-scid mice. Diabetologia 48:75-82
130. van Hensbergen Y, Schipper LF, Brand A, Slot MC, Welling M, Nauta AJ and Fibbe WE (2006) Ex vivo culture of human CD34+ cord blood cells with thrombopoietin (TPO) accelerates platelet engraftment in a NOD/SCID mouse model. Exp Hematol 34:943-50
131. Verstegen MM, van Hennik PB, Terpstra W, van den Bos C, Wielenga JJ, van Rooijen N, Ploemacher RE, Wagemaker G and Wognum AW (1998) Transplantation of human umbilical cord blood cells in macrophage-depleted SCID mice: evidence for accessory cell involvement in expansion of immature CD34+CD38− cells. Blood 91:1966-76
132. Wahid S, Blades MC, De Lord D, Brown I, Blake G, Yanni G, Haskard DO, Panayi GS and Pitzalis C (2000) Tumour necrosis factor-alpha (TNF-alpha) enhances lymphocyte migration into rheumatoid synovial tissue transplanted into severe combined immunodeficient (SCID) mice. Clin Exp Immunol 122:133-42
133. Wang J, Kimura T, Asada R, Harada S, Yokota S, Kawamoto Y, Fujimura Y, Tsuji T, Ikehara S and Sonoda Y (2003) SCID-repopulating cell activity of human cord blood-derived CD34− cells assured by intra-bone marrow injection. Blood 101:2924-31
134. Watanabe S, Terashima K, Ohta S, Horibata S, Yajima M, Shiozawa Y, Dewan MZ, Yu Z, Ito M, Morio T, Shimizu N, Honda M and Yamamoto N (2007) Hematopoietic stem cell-engrafted NOD/SCID/IL2Rγnull mice develop human lymphoid system and induce long-lasting HIV-1 infection with specific humoral immune responses. Blood 109:212-8
135. Westerink MA, Metzger DW, Hutchins WA, Adkins AR, Holder PF, Pais LB, Gheesling LL and Carlone GM (1997) Primary human immune response to *Neisseria meningitidis* serogroup C in interleukin-12-treated severe combined immunodeficient mice engrafted with human peripheral blood lymphocytes. J Infect Dis 175:84-90
136. Woodland RT and Schmidt MR (2005) Homeostatic proliferation of B cells. Semin Immunol 17:209-17
137. Woods A, Chen HY, Trumbauer ME, Sirotina A, Cummings R and Zaller DM (1994) Human major histocompatibility complex class II-restricted T cell responses in transgenic mice. J Exp Med 180:173-81
138. Xia L, McDaniel JM, Yago T, Doeden A and McEver RP (2004) Surface fucosylation of human cord blood cells augments binding to P-selectin and E-selectin and enhances engraftment in bone marrow. Blood 104:3091-6
139. Yahata T, Ando K, Nakamura Y, Ueyama Y, Shimamura K, Tamaoki N, Kato S and Hotta T (2002) Functional human T lymphocyte development from cord blood CD34+ cells in non-obese diabetic/Shi-scid, IL-2 receptor gamma null mice. J Immunol 169:204-9
140. Ishikawa F, Yoshida S, Saito Y, Hijikata A, Kitamura H, Tanaka S, Nakamura R, Tanaka T, Tomiyama H, Saito N, Fukata M, Miyamoto T, Lyons B, Ohshima K, Uchida N, Taniguchi S, Ohara O, Akashi K, Harada M, Shultz LD (2007) Chemotherapy-resistant human AML stem cells home to and engraft within the bone-marrow endosteal region. Nat Biotechnol 25:1315-21

141. Zheng Y, Watanabe N, Nagamura-Inoue T, Igura K, Nagayama H, Tojo A, Tanosaki R, Takaue Y, Okamoto S and Takahashi TA (2003) Ex vivo manipulation of umbilical cord blood-derived hematopoietic stem/progenitor cells with recombinant human stem cell factor can up-regulate levels of homing-essential molecules to increase their transmigratory potential. Exp Hematol 31:1237-1246
142. Zheng Y, Sun A and Han ZC (2005) Stem cell factor improves SCID-repopulating activity of human umbilical cord blood-derived hematopoietic stem/progenitor cells in xenotransplanted NOD/SCID mouse model. Bone Marrow Transplant 35:137-42

NOD/Shi-*scid* IL2rγ^{null} (NOG) Mice More Appropriate for Humanized Mouse Models

M. Ito(✉), K. Kobayashi, and T. Nakahata

1 Introduction 54
2 Development of Complex Immunodeficient Mice in Japan 55
   2.1 NOD/Shi-*scid* IL2rγ^{null} Mice 55
   2.2 Other IL2rγ^{null} Combined Immunodeficient Mice 57
3 NOG Mice 58
   3.1 Immunodeficiencies in NOG Mice 58
   3.2 Gene Expression 58
   3.3 Other Characteristics 60
4 Humanized Mice Based on NOG Mice 61
   4.1 Hematopoietic Cells 61
   4.2 Cancer 65
   4.3 Others 66
5 Further Improvement of Immunodeficient Mice 67
   5.1 Introduction of Human Growth or Differentiation Factors 67
   5.2 Depletion of Cells from Immunodeficient Mice 68
   5.3 Introduction of HLA Genes 68
6 Conclusion 69
References 69

Abstract "Humanized mice," in which various kinds of human cells and tissues can be engrafted and retain the same functions as in humans, are extremely useful because human diseases can be studied directly. Using the newly combined immunodeficient NOD-*scid* IL2rγ^{null} mice and *Rag2*^{null} IL2rγ^{null} humanized mice, it has became possible to expand applications because various hematopoietic cells can be differentiated by human hematopoietic stem cell transplantation, and the

---

M. Ito
Laboratory of Immunology, Central Institute for Experimental Animals, 1430 Nogawa, Miyamae, Kawasaki 216-0001, Japan
mito@ciea.or.jp

human immune system can be reconstituted to some degree. This work has attracted attention worldwide, but the development and use of immunodeficient mice in Japan are not very well known or understood. This review describes the history and characteristics of the NOD/Shi-*scid IL2rγ^{null}* (NOG) and BALB/cA-*Rag2^{null} IL2rγ^{null}* mice that were established in Japan, including our unpublished data from researchers who are currently using these mice. In addition, we also describe the potential development of new immunodeficient mice that can be used as humanized mice in the future.

Abbreviations Asialo GM1: asialoganglioside gangliotetraosylceramide; BM: bone marrow; CB: cord blood; GM-CSF: granulocyte-macrophage colony-stimulating factor; HSC: hematopoietic stem cell; IFN-γ: interferon gamma; IL: interleukin; IL-2Rγ: interleukin 2 receptor gamma chain; MHC: major histocompatibility complex; MSC: mesenchymal stem cell; NK: natural killer; PBMC: peripheral blood mononuclear cell; RNAi: RNA interference; SCF: stem cell factor; TCR: T cell receptor

1 Introduction

Clarification of the genetic base that covers the entire human and mouse genetic makeup [55, 107] has helped to accelerate research on human diseases and potential therapeutic applications based on known genetic information. In addition, the search for genes related to diseases, their expression patterns, the interaction between genes, and functional analysis of the gene products can now be pursued more quickly over a much wider spectrum. The results obtained from these analyses will be reflected in gene therapy and therapeutic developments and in future preventive medicine. Since a human being or a human pathogen becomes the target, the testing of such medicines requires the involvement of a human subject or an in vivo experimental model that is highly similar to a human. Therefore, the clearer the gene and product functions become in humans, the more the interspecies gap will become an issue, and one of the means of overcoming the interspecies gap is to use laboratory animals in which human cells and tissues are engrafted and function as in humans.

Over a long period of time, attempts have been made to develop laboratory animals with engrafted human cells, tissues, and organs. With each of these attempts, there has been a history of improvement and development of better immunodeficient animals. Historically, it started with the discovery by Isaacson et al. [39] of nude (formally *Foxn1^{nu}*) mice that have no thymus and Beige (*Lyst^{bg}*) mice [49], XID (*Btk^{xid}*) mice [5], and their combinations [46, 74]. Because of the lack of a thymus in these nude mice, T cells do not develop and thus it is possible to implant human tumors. Therefore these mice have been widely used in research on anticancer drugs against human tumors [28]. However, the engraftment of normal human cells and tissues in nude mice was not as successful as hoped for. The breakthrough in overcoming the limitations of these mice came from Bosma et al. in 1983 [8] with their discovery of CB-17-*scid* (formally *Prkdc* ^{scid}) mice. These mice exhibit severe immunodeficiencies since they have neither functional T nor B cells. Since a similar mutation causes pediatricsevere combined immunodeficiency (SCID) syndrome, the mice were named after the *scid* mutation. Subsequently, the immunodeficiency was found to be generated because of mutation of the DNA-dependent protein kinase (*Prkdc*) genes that are involved in the double-chain DNA restoration and in the V(D)J rearrangements of TCR and immunoglobulin [52, 67]. In this mouse, normal human hematopoietic cells could be developed, which is difficult with nude mice [64, 72]. Thus, in order to improve the engraftment level of human hematopoietic cells in these mice, SCID congenic mice based on the genetic background of other inbred mice combined with spontaneous mutation or transgenic mice were developed, allowing for better engraftment of human cells and tissues in this mouse line [6, 14, 36, 68]. Through these attempts, it was found that engraftment levels for human cells were very high in the NOD-*scid* mice, which were established by introducing the *scid* gene into the nonobese diabetes model of NOD inbred mice [53, 57, 82, 104]; It was then reported that the high engraftment level observed in these mice was due to the multiple immunological disorders of the background NOD mice strain, that is, it was due to the reduction of hemolytic complement activity, macrophage function, and NK activity [93]. These mice have been used for about 10 years. Beginning in the first half of the 1990s these mice became the "gold standard" for xenotransplantation involving human cells, especially for research into the differentiation of the human hematopoietic cells from stem cells, which resulted in extensive progress being made [31, 63]. Further attempts have also been undertaken to develop mice that show higher engraftment levels than the NOD-*scid* mice developed through combinations with other immunodeficient mice. Further investigations determined that the engraftment level of human cells in the recently developed NOD-*scid IL2rγ^{null}* mice and *Rag2^{null} IL2rγ^{null}* mice when combined with *IL2rγ^{null}* mice was extremely high, and that greater differentiation of various hematopoietic cells occurred after human stem cell transplantation compared with the conventional NOD-*scid* mice [41, 92, 103]. Various human cells differentiated from hematopoietic stem cells (HSCs) in these mice will result in successful reconstitution of the human immune system. Therefore, these mice are considered to be closer to true "humanized mice" at the present time.

2 Development of Complex Immunodeficient Mice in Japan

2.1 NOD/*Shi*-scid IL2rγ ^{null} Mice

A schematic diagram of the development of NOD/Shi-*scid IL2rγ^{null}* (NOG) mice in Japan is shown in Fig. 1. The development of various immunodeficient mice worldwide was reviewed by Shultz et al. [91] with the formal naming of them according

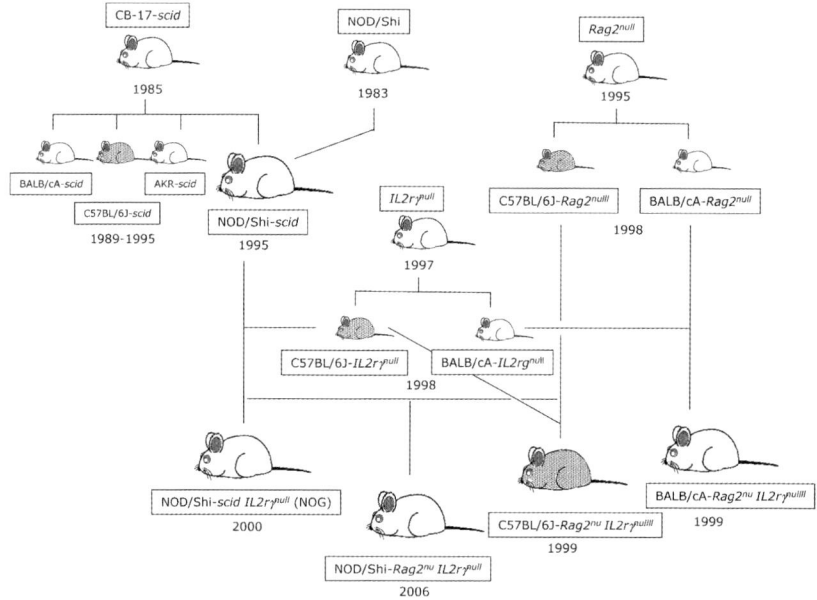

Fig. 1 Schematic diagram of the development of various immunodeficient mice in Japan. Strain nomenclature: CB-17-*scid*, C.Bka*Igh*^{b}-*Prkdc* ^{scid}/IcrJic; NOD-*scid*, NOD.CB17-*Prkdc* ^{scid}/ShiJic; BALB/cA-*Rag2*^{null}, C.129S1-*Rag2*^{tm1Fwa}/AJic; C57/B6J-*Rag2*^{null}, B6.129S1-*Rag2*^{tm1Fwa}/JJic; BALB/cA-*IL2rγ*^{null}, C.129S1-*Il2rg*^{tm1Sug}/ShiJic; C57/B6J-*IL2rγ*^{null}, B6.129S1-*Il2rg*^{tm1Sug}/Jic; NOG (NOD/Shi-*scid IL2rγ*^{null}), NOD.Cg-*Prkdc*^{scid} *Il2rg*^{tm1Sug}/ShiJic; BALB/cA-*Rag2*^{null} *IL2rγ*^{nul}, C.Cg-*Rag2*^{tm1Fwa} *Il2rg*^{tm1Sug}/AJic; C57B6J-*Rag2*^{null} *IL2rγ*^{nul}, B6.Cg-*Rag2*^{tm1Fwa} *Il2rg*^{tm1Sug}/JJic; NOD/Shi-*Rag2*^{null} *IL2rγ*^{nul}, NOD.Cg-*Rag2*^{tm1Fwa} *Il2rg*^{tm1Sug}/ShiJic

to International Committee on Standardized Genetic Nomenclature for Mice. NOG mice were established based on NOD/Shi-*scid* mice, one of the SCID congenic strains developed by the Central Institute for Experimental Animals (CIEA) [53]. Based on the NOD/LtSz-*scid* mice, Shultz et al. at the Jackson Laboratory developed NOD/LtSz-*scid IL2rγ*^{null} mice. Our original purpose for developing this mouse was not to improve the engraftment level, but to clarify the role of the subpopulation of T and B cells in the development of nonobese diabetes, which occurred in autoimmune NOD mice. Afterwards, it became clear that this strain of mice exhibited a better engraftment level for xenotransplants, including human hematopoietic cells compared to the other SCID congenic mice, and thus they have become widely used for xenotransplantation. However, differentiation of single CD4$^+$ and CD8$^+$ T cells have rarely been observed even in these mice. In these NOD-*scid* mice, the elimination of NK activity or NK cells by anti-NK antibody, anti-asialo GM1 antibody [32, 48], and anti-IL-2Rβ antibody Tmβ2 [100] treatment has been found to augment the engraftment of human cells [115]. This finding has led to the generation of new immunodeficient mice in which NK cells or NK activities are abolished.

At the time that genetic manipulation of the NOD-*scid* mice was being attempted, it was well known that *b2m*^{null} mice [116], *Perforin*^{null} mice [44], and

Granzyme B^{null} mice [89] have no or impared NK activity and *IL2rγ*^{null} mice [11, 22, 42, 78] and *IL2rβ* transgenic (Tg) mice [98] lack NK cells. Introducing these targeted genes into the NOD-*scid* mice led to the development of an improved NOD-*scid* mouse. The Shultz group developed NOD/LtSz-*scid b2m*^{null} mice [13], the related NOD/LtSz-*Rag1*^{null} *Pfn*^{null} mice [89], and more recently NOD/LtSz-*scid IL2rγ*^{null} mice [92]. We also attempted to improve NOD/Shi-*scid* mice by backcrossing the latter two mice, since this would result in NOD-*scid* mice that lacked NK cells. Among these two mouse lines, the development of NOD/Shi-*scid* mice with the IL2rβ transgene was stopped midway because of their low production efficiency. In 2000, NOD/Shi-*scid IL2rγ*^{null} (NOG) mice with the *IL2rγ*^{null} gene were successfully established at last and were able to be maintained with good production efficiency. Subsequently, as was reported in 2002 [41], experimental studies with human hematopoietic stem cells transferred to these mice demonstrated that they were extremely efficient for humanized mice. NOG mice were established by a 10th generation backcrossing of *IL2rγ*^{null} mice to NOD/Shi-*scid* mice. There are two differences between Shultz's NOD-*scid IL2rγ*^{null} mice and our mouse line. These differences are related to the background substrain of the NOD mice and the *IL2rγ*^{null} mice used for backcrossing during development of the NOD-*scid* mice. However, the difference in multiple immunological disorders between NOD/Shi-*scid* and NOD/LtSz-*scid* mice has yet to be found [41]. The *IL2rγ*^{null} mice that we used were generated by Ohbo et al. [78]. In these mice, exon 7 was targeted and there was a truncated form of IL-2Rγ without intracellular signaling, which could be stained with an anti-IL-2Rγ antibody. The *IL2rγ*^{null} mice of Cao et al. [11] were used by Shultz et al., and in these mice, exon 1 was targeted. Therefore, the formal names for the NOD-*scid IL2rγ*^{null} mice are NOD.Cg-*Prkdc*^{scid}*Il2rg*^{tm1Sug}/Jic and NOD.Cg-*Prkdc*^{scid}*Il2rg*^{tm1Wjl}/SzJ, respectively. We named our mice after our "NOG mice." To date, the difference in engraftment level is not completely clear, because engraftment efficiency has never been compared between these mice, although it is considered to be basically the same.

2.2 Other IL2rγ^{null} Combined Immunodeficient Mice

In addition to the development of NOG mice, another type of immunodeficient mice replacing the *Rag2*^{null} genes for the *scid* mutation has been developed, because these inactive genes cause the same phenotypic T and B cell deficiency in the mice. In 1998, Goldman et al. were the first to report this combination mouse [29], in which the proliferation of B-lymphoblastoid cells and engraftment rate for human peripheral blood lymphocytes (PBL) were higher than those seen for NOD/LtSz-*scid* mice. The genetic backgrounds were a mixture of 129Ola, BALB/c, and C57BL/6 mice. Afterwards, Kirberg et al. reported on H2^d- *Rag2*^{null} *IL2rγ*^{null} mice [51], but the genetic background of the strain was outbred. We have also developed *Rag2*^{null} *IL2rγ*^{null} mice that have genetic backgrounds that consist of respective BALB/cA, C57BL/6J, or NOD/Shi inbred strains (Fig. 1). Dr. Shultz and his coworkers also described NOD/LtSz-*Rag1*^{null} *IL2rγ*^{null} in their review [91].

The NOG mice that we developed were mainly used in Japan for various research studies on xenotransplantation, including human hematopoiesis. BALB/cA and C57BL/6J-*Rag2*^{null} *IL2rγ*^{null} mice were sent to the United States and later on to Europe, with the successful development of humanized mice that used irradiated newborn BALB/c-*Rag2*^{null} *IL2rγ*^{null} mice, as reported by Traggiai et al. in 2003 [103].

3 NOG Mice

3.1 *Immunodeficiencies in NOG Mice*

NOG mice have multiple immunodeficiencies that are principally derived from three strains of mice. These include:

1. Reduced innate immunity derived from a NOD inbred strain, which involves a macrophage dysfunction, and a defect of complement hemolytic activity and reduced NK activity [93]. The NOD/Shi inbred strain was first discovered by Makino et al. as autoimmune non-obese-type diabetes mice [50, 58].
2. Lack of functional T and B cells that is derived from a mutation of protein kinase (*Prkdc*: protein kinase, DNA activated, catalytic polypeptide), which is the causative gene of the *scid* mutation [8, 52].
3. Lack of NK cells, dendritic cell dysfunctions, and other unknown deficiencies due to inactivation of the IL-2Rγ gene [38, 78, 79, 97].

When transfer of human umbilical cord blood CD34$^+$ cells to NOG mice was performed, growth and differentiation of various hematopoietic cells were observed. This high level of engraftment and differentiation in NOG mice cannot be traced only to elimination of the NK cells from the NOD-*scid* mice. In fact, remarkable differences (3- to 4-fold) in the engraftment rate and differentiation of human hematopoietic cells have been observed between NOG and NOD/LtSz-*scid b2m*^{null} mice that lack NK activity [41]. In addition, the engraftment rates in NOG mice appear to be relatively uniform, although in some of the NOD-*scid* mice it was noted that engraftment often failed to occur. The unknown factors that might be responsible for supporting the higher engraftment have yet to be elucidated. One of these factors may be the lack of interferon gamma (IFN-γ) production in dendritic cells of NOG mice as described below.

3.2 *Gene Expression*

To determine the factors responsible for higher engraftment of human cells in NOG mice, the expression of genes related to the immunological responses in spleen cells were compared among the different NOG mice, NOD/Shi-*scid* mice treated with anti-asialo GM1 antibodies to eliminate NK cells, and NOD/Shi mice. As seen in Fig. 2, there was a remarkable reduction in the gene expression of IFN-γ, and in IP10, Mig, and RANTES that was secondarily induced by IFN-γ [108]. Proinflammatory cytokine IFN-γ is well known as an important factor that is responsible for various cell signal transductions [1, 7]. A recent report on IFN-γ inducing killer dendritic cells that found that these cells play a role in the rejection of xenotransplants [12, 99] suggests that lack of IFN-γ may be a critical factor for supporting higher engraftment of xenotransplants in NOG mice. This is consistent with our unpublished data showing that NOD-*scid*-derived CD11c+ spleen cell transfer into NOG mice reduced the engraftment of human PBL in NOG mice. Clarification of this issue may lead to discovery of new xenotransplant rejection mechanisms and, moreover, lead to further improvement of immunodeficient mice.

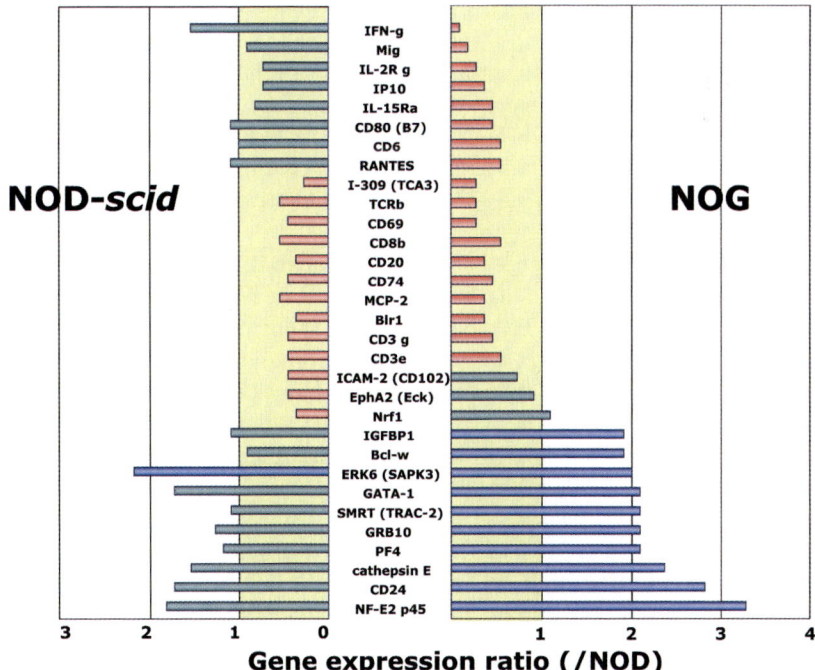

Fig. 2 Expression of the genes associated with immunological responses in NOD, NOD/Shi-*scid* and NOG mice. At 48 h after intraperitoneal infection of 1×10⁷ *Listeria monocytogenes*, spleens were removed and the RNA was extracted. Gene expression was examined by microarray (Toyobo Inc.). The *yellow areas* express the intensity of gene expression in the NOD mice. The *blue bars* express higher intensity, and the *gray bars* express lower intensity when compared with those of the NOD/Shi mice

3.3 Other Characteristics

Although NOG mice are extremely immunodeficient, when kept under strict SPF conditions, NOG mice can live for more than one and a half years, which is the same life span as that of conventional inbred mice. NOD/LtSz-*scid IL2rγ^{null}* mice also can survive over one and a half years. However, half of the NOD/LtSz-*scid* mice die within one year [90]. The body weight change of NOG mice was almost the same as that seen for the NOD/Shi-*scid* mice, with weights of 23.0 ± 2.0 g for males and 19.7 ± 1.3 g for females at 8 weeks of age, and 28.1 ± 1.1 g for males and 23.0 ± 1.2 g for females at 16 weeks of ages. Lymphoid organs in both the NOG and BALB/cA-*Rag2^{null} IL2rγ^{null}* mice were remarkably immature and exhibited atrophy. The average spleen weight (23 mg for females and 24 mg for males) for NOG mice was less than that noted for the NOD/Shi-*scid* mice (27 mg for females and 28 mg for males). The thymus of the NOG mice was also smaller than that of NOD/Shi-*scid* mice and can be detected at a rudimental level by histological observation. The lymph nodes could not be seen macroscopically.

Rag1, *Rag2*, and *Prkdc* (a causative gene of the *scid* mutation) are indispensable genes for the rearrangement of TCR and immunoglobulin, with the inactivation of these genes in mice resulting in a similar phenotype that lacks both T and B cells [25, 70, 88]. *Prkdc* genes also act as DNA repair machinery. Mice with the *scid* mutation have additional phenotypes, such as irradiation sensitivity, high occurrence of thymic lymphoma, and leakiness, which has been described as a phenomenon where immunoglobulin and aberrant T cells occur in aging CB-17-*scid* mice. Thus, there is a possibility that NOG mice will show these phenomena simply because they have the *scid* mutation gene.

Irradiation sensitivity is higher in NOG mice than in BALB/cA-*Rag2^{null} IL2rγ^{null}* mice. In our studies, we found that when mice at 10 weeks of age were exposed to whole body irradiation with 3 Gy, about one-third of the irradiated mice were killed by 3 weeks. However, this was not seen when the irradiation level was 2.5 Gy. In contrast, NOD/LtSz-*scid IL2rγ^{null}* mice survived in irradiation up to levels of 3.5 Gy [92]. This difference may be associated with the radiation source and the radiation dose per minute. On the other hand, for BALB/cA-*Rag2^{null} IL2rγ^{null}* mice with the *Rag2^{null}* for *scid* mutation, the quantity of irradiation used can be higher than that used in NOG mice because of their X-ray resistance [27, 103]. Whole body irradiation in newborn NOD-*scid IL2rγ^{null}* and BALB/cA-*Rag2^{null} IL2rγ^{null}* mice was performed with 1 Gy and 3.5 Gy, respectively.

It is well known that thymic lymphoma occurs frequently in CB-17-*scid* mice [15]. From our unpublished data, we found that thymoma frequently occurred 30-50 weeks after birth for CB-17-*scid* mice, and thymomas were observed in 69 (about 27%) of 255 mice during the 2 years of observation. A much higher incidence of thymoma has been reported in NOD/LtSz-*scid* mice [83]. In NOD/Shi-*scid* mice, 18.4% had a thymus tumor at 10 months of age, which is remarkably high when compared with the 9% incidence seen in the CB-17-*scid* mice at 10 months of age (unpublished data). On the other hand, thymomas are rare in NOG mice. To date, thymomas have been observed in only 4 of more than 5,000 mice, including mice with transfer of various human cells, including human umbilical cord blood CD34+ cells. The disadvantage of NOD-*scid IL2rγ*^{null} mice has generally been considered to be the high incidence of thymoma as compared to BALB/cA-*Rag2*^{null} *IL2rγ*^{null} mice [4, 103]. However, this could be a misunderstanding, because the reasons why thymomas are so frequently observed in NOD-*scid* mice and why they do not occur in NOG mice are really not all that clear. The reason might be a lack of factors such as IL-2, IL-4, and IL-7, which are responsible for the growth of T cells by inactivation of the IL-2R γ gene [97].

Another feature of mice with the *scid* gene involves the incidence of T and B cells that are associated with aging, which is referred to as "leakiness" [9]. For CB-17-*scid* mice, more than 90% of the mice begin to produce more than 1 μg/ml of immunoglobulin in the serum by one and a half years after birth. However, when compared with the leakiness observed in CB-17-*scid* mice, little is found in NOD/Shi-*scid* mice, with no leakiness at all recognized in NOG mice. The reason for this remains to be elucidated.

4 Humanized Mice Based on NOG Mice

4.1 *Hematopoietic Cells*

Many studies have been performed using NOD-*scid* mice to examine differentiation of hematopoietic stem cells (HSCs) [20, 21]. The most attractive characteristic of NOD-*scid IL2rγ*^{null} mice in this research field is that various human hematopoietic cells develop with high engraftment and have been shown to be able to survive for a long time with the transfer of human umbilical cord blood CD34+ stem cells. Thus, the human immune system can be reconstructed in these mice. In Table 1, the results obtained to date using NOD -*scid IL2rγ*^{null} and *Rag2*^{null} *Il2rγ*^{null} mice are summarized.

The degree of differentiation and growth of human hematopoietic cells in these mice via transfer of HSCs varies depending upon the donor cells, cell number, mouse age at transfer, and days after transfer (Table 2). Constant detection of differentiated human cells in various organs was possible when at least 5×10^4 CB CD34+ cells were transferred. Human cells could be engrafted in 2 of 6 mice in which only 100 CD34+ cells were transferred [41]. As a source of inoculated stem cells, CD34+ cells derived from cord blood seem to be more efficient for engraftment, although all CD34+ cells from cord blood, bone marrow, and peripheral blood can successfully grow and differentiate in NOG mice [61]. Three serial passages of engrafted CD34+ cells in NOG mice to a different NOG mouse in bone marrow of these mice was found to be possible [Dr. Kiyoshi Ando of Tokai University, personal communication].

Table 1 Results of human stem cell transfer experiments using NOD–*scid IL2rγ*[null] and *Rag2*[null] *IL2rγ*[null] mice

	Mouse strain	Adult or newborn	Inoculation route	Cell source	Results	References
1	NOG	Adult/240 cGy	Tail vein	CB CD34+ 1×10^5, 4×10^4	First report on NOG mice, 15.9%* in PB** (2 months)	Ito et al. (2002)
2	NOG	Adult/250 cGy/ 7–9 weeks old	Tail vein	CB CD34+ 8–20×10^4	54.6% in PB, 72.9% in BM (over 19 weeks)/6.8% in PB, 41.9% in BM of NOD–*scid*	Yahata et al. (2002)
4	NOG	Adult/240 cGy/ 8–10 weeks old	Tail vein	CB CD34+ 1×10^4, 5×10^4	72.6% in SP (4 months)/functional T, B, and NK cells/human cell reconstitution in organs	Hiramatsu et al. (2003)
5	NOG	Adult/250 cGy/ 9 weeks old	Tail vein	hCB CD34+ 1.8–5.6×10^5/hBM CD34+ 1.9–14×10^5/hMPB CD34+/ 5–18×10^5	CB: 47.5% in PB, 66.7% in BM. 63.7% in SP/BM: 8.1% in PB, 22.5% in BM, 14.0% in SP/MPB: 10.7% in PB, 45.4% in BM, 13.8% in SP (14 weeks)	Matsumura et al. (2003)
6	NOG	Adult/300 cGy/6–10 weeks old	Tail vein	CB CD34+ 1×10^4–1.2×10^5	18.8% in PB, 43.9% in SP (2 months)/26.1% in PB, 52.3% in SP/67.7% in PB, 81.9% in SP/lymphoid–like structure in spleen, R–5, X–4 HIV-1 infection, production of anti–HIV–1 antibodies	Watanabe et al. (2007)
7	NOD/LtSz–*scid IL2rγ*[null]	Newborn/100 cGy	Facial vein	CB CD34+ 1×10^5, CB CD34+, CD38–2 x 10^4	68.9% in PB, 72.9% in BM, 54.5% in SP (3 months)/OVA–specific IgM and IgA+/erythrocytes+, DCs+, other progenitors+	Ishikawa et al. (2005)
8	NOD/LtSz–*scid IL2rγ*[null]	Adult/ 325 cGy	Tail vein	CB CD34+ 7×10^5	34.9% in BM, 59.9% in SP, 36.6% in THY, 6.3% in PB (10 weeks)/78.3% in THY, 20.7% in PB (+FC–IL–7)	Shultz et al. (2005)
9	H2D– *Rag2*[null] *IL2rγ*[null]	Adult/350 cGy	Tail vein	CB CD34+ 5×10^5	33% in BM, 4% in PB (7 weeks)	Rozenmuller et al. (2004)

10	BALB/cA–$Rag2^{null}IL2r\gamma^{null}$	Newborn/ 2×200 cGy (3- to 4-h interval) 375 cGy/min	Intrahepatically	CB CD34$^+$ 3.8-12×10^4		5% to ~70% in BM, 0 to ~83% in BM/positive EB infection, anti-TT IgG formation, white pulplike structure in spleen	Traggiai et al. (2004)
11	BALB/cA–$Rag2^{null}IL2r\gamma^{null}$	Newborn/350 cGy	Intrahepatically	CB CD34$^+$ 1×10^6		5%–89% in SP (12 weeks)/HIV-1 infection	Berges et al. (2006)

* % of human CD45$^+$ cells
** PB: peripheral blood, BM: bone marrow, SP: spleen, THY: thymus

Table 2 Cell differentiation after human HSC transfer in NOG mice

Cell	Development*
Lymphocytes	++
T cells	+++
B cells	++
NK cells	+
Macrophages	+
Dendritic cells	+
Mast cells	+
Monocytes	+
Neutrophils	– to +
Platelets	+
Erythrocytes	– to +

*12–16 weeks after human HSC transfer

After human stem cell transfer to NOG mice, the human (h) CD19$^+$ B cell differentiates first. Most hCD45$^+$ cells in the peripheral blood are B cells by about 6-10 weeks after cell transfer. T cells will differentiate later, around 8-16 weeks. hCD4 single positive and hCD8 single positive T cells, which are rarely observed in NOD-*scid* mice, can be detected in the peripheral blood and spleen. At the present time, T cells are considered to develop in the mouse thymus. In the thymus, hCD4hCD8 double-positive cells reached 60% to 75%. In contrast, hCD4$^+$ T cells ranged from 40% to 70% and hCD8$^+$ cells ranged from 30% to 50% in hCD45$^+$ cells of the spleen and but a few hCD4hCD8 double-positive cells were found [37, 61, 112]. These observations are consistent with known T cell development in human organs. These T cells show diversity of the TCR, which is activated by stimulation with anti-human CD3 antibodies, indicating that T cells that develop in NOG mice are functionally active. B cells express IgM and IgD on the surface and can produce antigen-specific IgM antibodies, but not a lot of antigen-specific IgG is produced even with multiple antigen challenges. Matsumura et al. [61] reported that CD5$^+$ B1 type B cells are well differentiated, but CD5$^-$ B2 type B cells were not, suggesting that CD5$^-$ B2 type B cells may be required to efficiently produce antigen-specific IgG. A lymphoid-like structure develops in the spleen and contains human macrophages and dendritic cells [106]. However, human follicular dendritic cells (FDC) have yet to be observed. Natural killer cells are also detected from 12 weeks in the peripheral blood and spleen after stem cell transfer [36]. Mast cells are recognized within the skin, lung, stomach, and intestine from 20 weeks after transfer, and connective-tissue type (tryptase$^+$/chymase$^+$) and mucosal type (tryptase$^+$/chymase$^-$) mast cells are localized and consistent with human organs [45]. While the differentiation of lymphoid lineage cells is well recognized, this is not the case for cells of the myeloid lineage. Levels of platelets that are detected in the peripheral blood at 8 weeks after CB CD34$^+$ cell transfer range from 0.2% to 3% [75]. Erythrocytes and neutrophils are not usually recognized in the peripheral blood. Human neutrophils in HSC-engrafted NOG mice can be detected by a zymosan-induced air pouch inflammation technique [23]. Nucleated red blood cells and megakaryocytes could be detected to some degree in mouse bone marrow transfer, indicating that the factors responsible for denucleation of these cells might not all be fully present in NOG mice. However, fully matured erythrocytes could be detected in the peripheral blood when HSCs were transferred to newborn NOD/LtSz-*scid, IL2rγ^{null}* mice [40]. It is possible that when using newborn mice, the degree of engraftment and differentiation could be accelerated, although direct experiments to compare newborns and adults have yet to be performed.

To date, there is no evidence on the relative merits of NOG and BALB/cA-*Rag2^{null} IL2rγ^{null}* mice for humanized mice. The reason for this is the differences in handling of these mice. In NOG mice, HSCs are usually transferred into irradiated adult mice via the tail vein. In contrast, in irradiated newborns they are transferred intrahepatically in BALB/cA-*Rag2^{null} IL2rγ^{null}* mice. Recent results using irradiated newborn NOD/LtSz-*scid IL2Rγ^{null}* mice by Ishikawa et al. [40] suggest that higher engraftment and differentiation of HSCs can be obtained by using newborns rather than adults. Preliminary intravenous transfer experiments using CB CD34$^+$ cells in irradiated adult mice in order to compare the engraftment rate and differentiation among NOG, BALB/cA-*Rag2^{null} IL2rγ^{null}*, C57BL/6J-*Rag2^{null} IL2rγ^{null}*, and NOD/Shi-*scid* mice have provided interesting results. When the engraftment rates were compared among these mice, they were found to be the highest in NOG mice, moderate in the NOD/Shi-*scid* and BALB/cA-*Rag2^{null} IL2rγ^{null}* mice, and extremely low in the C57BL/6J-*Rag2^{null} IL2rγ^{null}* mice. Interestingly, hCD3$^+$, hCD4$^+$, and hCD8$^+$ T cells were highly developed in the NOG and BALB/cA-*Rag2^{null} IL2rγ^{null}* mice, but not in NOD/Shi-*scid* and C57BL/6J-*Rag2^{null} IL2rγ^{null}* mice. Further experiments are needed to clearly determine which mice are more appropriate and how transfers should be undertaken [Dr. Kiyoshi Ando of Tokai University, personal communication].

It is known that the genetic background of inbred strains of mice can influence the engraftment of tumor cells [56]. The difference in engraftment between *Rag2^{null}* mice and SCID mice has also been reported [94]. These findings suggest that the combination of inactive genes and the genetic backgrounds of mouse strains are very important when selecting more appropriate mice for attempts to improve immunodeficient mice for a humanized mouse model. Selection of the transfer route for stem cells into immunodeficient mice may also be important from the viewpoint of homing of the transferred cells [81].

4.2 Cancer

In the field of cancer research, since the discovery of nude mice there has been a long history of engraftment of a variety of tumor cells, including primary cells and tumor cell lines [28]. A liver metastasis model [96], a multiple myeloma model [19,

69], an acute myeloid leukemia model [77], and a Hodgkin lymphoma model [18] using NOG mice have all been reported. In the liver metastasis model, human pancreatic cancer cells, which were delivered by intrasplenic injection, highly metastasized in the liver in NOG mice but not in NOD/Shi-*scid* mice. In addition, the cells with metastatic capability were easily concentrated, allowing for identification of the genes associated with the metastasis. In the multiple myeloma model, intravenous transfer in NOG mice resulted in the development of multiple myeloma that resembled human myeloma. In these mice, the myeloma only grew in bone marrow, leading to paralysis in the hind legs. These observations suggest that tumor cells can grow when supported by mouse adhesion molecules, and in such cases, cell engraftment should be extremely high.

The identification of cancer stem cells is now one of the most studied areas in the field [80, 105]. Since tumor cells can easily grow in NOG mice after inoculation of just a small number of tumor cells, these mice can be very useful in this research.

4.3 Others

In the field of human infectious diseases, extensive research on the HIV-1 infection model is being undertaken. The hu-PBL-SCID mouse model, which can be infected with HIV-1 via transfer of human PBL, is well known [71, 102]. This model uses NOG mice and is also useful in evaluating anti-HIV-1 drugs, because of the high engraftment rate of transferred PBL [76]. However, this model develops a severe graft vs. host reaction that results in death by 2-3 months after cell transfer. Recently, a new mouse model with various human hematopoietic cells including CD4+ T cells has been developed through the transfer of CB CD34+ using NOG [106] and *Rag2^{null} Il2rγ^{nu}*mice [2, 4, 30]. In this model, HIV-1 infection persists for a long time without any graft vs. host reaction and with human immunity, and therefore research on vaccine development is now possible. The usefulness of NOG mice as Epstein-Barr virus (EBV) and human T-cell leukemia virus type 1 (HLTV-1) infection models has also been reported [16, 17]. So far there have been no reports concerning hepatitis C virus (HCV) infection studies using this model. Human albumin- and α-anti-trypsin-positive cells have been detected after transfer of CD34+ cells in NOG mice, although these cells were also positive for both human and mouse MHC, which resulted from human CD3+ cells that fused with mouse liver cells [26]. Studies on replacement of mouse liver cells with human liver cells have been performed in another mouse model [66, 101]. By combining with NOD-*scid, IL2rγ^{null}*, or *Rag2^{null} IL2rγ^{null}* mice, more useful models with human liver and the human immune system will become available. Other studies using NOG mice included reconstitution of human endometrium after implantation of endometrial tissue or cells resulting in induction of a menstrual cycle dependent on human sexual hormone [60, 62], and drug evaluation for human thrombopoiesis [75]. Retroviral introduction of a green fluorescent protein gene into HSCs can be used to perform research on the distribution or homing of transferred human cells [111, 113].

5 Further Improvement of Immunodeficient Mice

There are two approaches to further improve immunodeficient mice for "humanized mice." The first requires investigation of the inoculant side and involves cell sources including cotransplantation with mesenchymal stem cells, bone marrow, fetal livers, and inoculation routes [65, 73], while the second involves improvement on the recipient mouse side. Here, we refer to the latter. The improvement of NOD-*scid IL2rγ*^{null} or BALB/cA-*Rag2*^{null} *IL2rγ*^{null} mice is a practical approach. For this to occur, there are three approaches that can be used with current transgenic technology. First, there must be introduction of human genes into the mice that code for growth or differentiation factors. Second, depletion of cells, that is, macrophages, dendritic cells, mast cells, neutrophils, etc., which are responsible for innate immunity and which still remain in NOG mice, or their differentiation factors must be performed. Third, introduction of HLA genes into the mice must be performed in order to facilitate cell-to-cell interactions. These steps have been already performed with NOD-*scid* mice.

Generation of transgenic mice that are based on NOD-*scid IL2rγ*^{null} and BALB/cA-*Rag2*^{null} *IL2rγ*^{null} mice is easy by microinjection directly into the pronuclear embryos of these mice. The introduction of targeted genes into these mice can now be rapidly completed within 1 year through the use of the speed congenic technique with multiplex PCR [59, 110] different from the conventional technique by over 7-10 backcrossings without genetic analysis to replace the background genes.

5.1 Introduction of Human Growth or Differentiation Factors

Numerous works have been published on the introduction of human genes and their products for the purpose of enhancing engraftment rate and differentiation of transferred human cells in immunodeficient mice, including CB-17-*scid*, NOD-*scid* mice [81]. By applying this technique to immunodeficient mice, further improvements may be possible. However, it is unclear which factors are appropriate for this purpose. The cross-reactivity of factors between mice and humans also may influence human cell engraftment and differentiation. Recent advances have allowed us to identify the gene responsible for self-renewal of stem cells [87, 114], and their introduction into mice may make it possible to develop new immunodeficient mice in which human stem cells can be maintained for a long time. To further differentiate and grow myeloid lineage cells, introduction of genes of GM-CSF, IL-3, erythropoietin, thrombopoietin, etc., may be efficient. The factors responsible for denucleation of erythroid progenitors or megakaryocytes may be necessary. The genes of lymphotoxin-related molecules and chemokines that are necessary for reconstruction of human secondary lymphoid organs in these mice may also be targets for introduction [24, 95]. Genes related to inflammatory responses or human diseases, namely, IL-4, IL-5, etc. for allergic reactions, are interesting in that they may be useful in developing the disease in humanized mice [109].

5.2 Depletion of Cells from Immunodeficient Mice

To improve immunodeficient mice, depletion of the cells remaining in the immunodeficient mice may be an alternative, with the expectation that there might be a higher engraftment rate even when only a small number of stem cells are transferred. NOG and BALB/cA-$Rag2^{null}$ $IL2r\gamma^{null}$ mice lack T, B, and NK cells and also have a variety of immune disorders. However, these mice still have macrophages, dendritic cells, mast cells, and neutrophils, which play a role in innate immunity. To deplete such cells, the transgenic technique using genes of the herpes simplex virus thymidine kinase, diphtheria toxin A, and more recently, the diphtheria toxin receptor will be useful [3, 34, 43, 47, 85]. The partial depletion of a particular cell with the RNAi technique through the suppression of specific gene expression to a certain cell may also be effective [84, 86].

5.3 Introduction of HLA Genes

The introduction or replacement of HLA genes in NOG and BALB/cA-$Rag2^{null}$ $IL2r\gamma^{null}$ mice is considered to be a potent method for humanized mice. Many studies using these techniques for HLA transgenic mice have also been performed [80]. Intrathymic expression of human HLA-DR1 in NOD-*scid* mice has clearly accelerated T cell engraftment and their responses in these mice [10]. Lower production of antigen-specific IgG is considered to be one of the current issues in humanized mice based on NOG mice. One of the reasons for this may be the lack of a cell-to-cell interaction between human T cells and B cells, or macrophages, since the T cells that develop in the mice are educated in the mouse thymus. Therefore, the introduction of HLA into the mice should provide a more complete form of human immune system in the mice. However, it is also well known that HLA shows extreme diversity and that human diseases are associated with a particular HLA type [35]. Therefore, the choice of the HLA type to introduce into the mice may be not only an important but also a difficult issue to deal with. A newly developed transgenic technology using artificial chromosomes may be of help in introducing multiple HLA type genes [54]. On the other hand, there is an unsolved issue with regard to current humanized mice based on NOG mice. The observation of HLA-dependent T cell cytotoxicity in HSC-transferred NOD/LtSz-*scid*, $IL2r\gamma^{null}$ mice [40] cannot rule out the existence of extrathymic development of human T cells in the humanized mice. This issue is both interesting and important to clarify.

It is possible to improve immunodeficient mice through the introduction of genes with current transgenic technology, as has been described above. However, although such multiple modifications in the mice may be theoretically possible, they will lower production efficacy of the mice. It will only be after many trials are undertaken to generate various types of immunodeficient mice that a new immunodeficient mouse more appropriate for use as a humanized mouse may eventually be established.

6 Conclusion

Humanized mice having human cells, tissues, and organs, which will facilitate the study of the mechanism of pathogenesis in human diseases and help develop medicines that can be used directly to treat such diseases, are extremely important and a highly desirable goal. For humanized mice, various immunodeficient mice have been established through the introduction of targeted genes in existing immunodeficient mice. Recently developed NOD-*scid IL2rγ^{null}* and *Rag2^{null} IL2rγ^{null}* mice, which have a *scid* mutation gene or a *Rag2^{null}* gene and an *IL2rγ^{null}* gene, have garnered wide attention because of high engraftment and differentiation rates of the human hematopoietic cells from stem cells that occur in these mice. Therefore, these mice are now considered to be the best for use as a humanized mouse. However, because of the limitations, these mice have not yet reached the optimal level, and the process should be accelerated so that the ultimate immunodeficient mouse can be developed.

Acknowledgements We thank Dr. Masataka Nakamura, Tokyo Medical and Dental University, and Dr. Yoshito Ueyama, Tokai University School of Medicine, for cooperation in generation of NOG and *Rag2^{null} IL2rγ^{null}* mice. This work was supported by the Ministry of Education, Culture, Sports, Science and Technology, Ministry of Health, Labor and Welfare, Japan Science and Technology Agency (JST), and Kanagawa Academy of Science and Technology (KAST).

References

1. Bach, E. A., M. Aguet, and R. D. Schreiber. 1997. The IFN gamma receptor: a paradigm for cytokine receptor signaling. *Annu Rev Immunol* 15:563-591.
2. Baenziger, S., R. Tussiwand, E. Schlaepfer, L. Mazzucchelli, M. Heikenwalder, M. O. Kurrer, S. Behnke, J. Frey, A. Oxenius, H. Joller, A. Aguzzi, M. G. Manz, and R. F. Speck. 2006. Disseminated and sustained HIV infection in CD34+ cord blood cell-transplanted Rag2$^{-/-}$gamma c$^{-/-}$ mice. *Proc Natl Acad Sci USA* 103:15951-15956.
3. Behringer, R. R., L. S. Mathews, R. D. Palmiter, and R. L. Brinster. 1988. Dwarf mice produced by genetic ablation of growth hormone-expressing cells. *Genes Dev* 2:453-461.

4. Berges, B. K., W. H. Wheat, B. E. Palmer, E. Connick, and R. Akkina. 2006. HIV-1 infection and CD4 T cell depletion in the humanized Rag2$^{-/-}$gamma c$^{-/-}$ (RAG-hu) mouse model. *Retrovirology* 3:76.
5. Berning, A. K., E. M. Eicher, W. E. Paul, and I. Scher. 1980. Mapping of the X-linked immune deficiency mutation (xid) of CBA/N mice. *J Immunol* 124:1875-1877.
6. Bock, T. A., D. Orlic, C. E. Dunbar, H. E. Broxmeyer, and D. M. Bodine. 1995. Improved engraftment of human hematopoietic cells in severe combined immunodeficient (SCID) mice carrying human cytokine transgenes. *J Exp Med* 182:2037-2043.
7. Boehm, U., T. Klamp, M. Groot, and J. C. Howard. 1997. Cellular responses to interferon-gamma. *Annu Rev Immunol* 15:749-795.
8. Bosma, G. C., R. P. Custer, and M. J. Bosma. 1983. A severe combined immunodeficiency mutation in the mouse. *Nature* 301:527-530.
9. Bosma, M. J. 1992. B and T cell leakiness in the scid mouse mutant. *Immunodeficiency Rev* 3:261-276.
10. Camacho, R. E., R. Wnek, K. Shah, D. M. Zaller, R. J. O'Reilly, N. Collins, P. Fitzgerald-Bocarsly, and G. C. Koo. 2004. Intra-thymic/splenic engraftment of human T cells in HLA-DR1 transgenic NOD/scid mice. *Cell Immunol* 232:86-95.
11. Cao, X., E. W. Shores, J. Hu-Li, M. R. Anver, B. L. Kelsall, S. M. Russell, J. Drago, M. Noguchi, A. Grinberg, E. T. Bloom et al. 1995. Defective lymphoid development in mice lacking expression of the common cytokine receptor gamma chain. *Immunity* 2:223-238.
12. Chan, C. W., E. Crafton, H. N. Fan, J. Flook, K. Yoshimura, M. Skarica, D. Brockstedt, T. W. Dubensky, M. F. Stins, L. L. Lanier, D. M. Pardoll, and F. Housseau. 2006. Interferon-producing killer dendritic cells provide a link between innate and adaptive immunity. *Nat Med* 12:207-213.
13. Christianson, S. W., D. L. Greiner, R. A. Hesselton, J. H. Leif, E. J. Wagar, I. B. Schweitzer, T. V. Rajan, B. Gott, D. C. Roopenian, and L. D. Shultz. 1997. Enhanced human CD4+ T cell engraftment in beta2-microglobulin-deficient NOD-scid mice. *J Immunol* 158:3578-3586.
14. Christianson, S. W., D. L. Greiner, I. B. Schweitzer, B. Gott, G. L. Beamer, P. A. Schweitzer, R. M. Hesselton, and L. D. Shultz. 1996. Role of natural killer cells on engraftment of human lymphoid cells and on metastasis of human T-lymphoblastoid leukemia cells in C57BL/6J-scid mice and in C57BL/6J-scid bg mice. *Cell immunol* 171:186-199.
15. Custer, R. P., G. C. Bosma, and M. J. Bosma. 1985. Severe combined immunodeficiency (SCID) in the mouse. Pathology, reconstitution, neoplasms. *Am J Pathol* 120:464-477.
16. Dewan, M. Z., K. Terashima, M. Taruishi, H. Hasegawa, M. Ito, Y. Tanaka, N. Mori, T. Sata, Y. Koyanagi, M. Maeda, Y. Kubuki, A. Okayama, M. Fujii, and N. Yamamoto. 2003. Rapid tumor formation of human T-cell leukemia virus type 1-infected cell lines in novel NOD-SCID/γc^{null} mice: suppression by an Inhibitor against NF-kappaB. *J Virol* 77:5286-5294.
17. Dewan, M. Z., J. N. Uchihara, K. Terashima, M. Honda, T. Sata, M. Ito, N. Fujii, K. Uozumi, K. Tsukasaki, M. Tomonaga, Y. Kubuki, A. Okayama, M. Toi, N. Mori, and N. Yamamoto. 2006. Efficient intervention of growth and infiltration of primary adult T-cell leukemia cells by an HIV protease inhibitor, ritonavir. *Blood* 107:716-724.
18. Dewan, M. Z., M. Watanabe, S. Ahmed, K. Terashima, S. Horiuchi, T. Sata, M. Honda, M. Ito, T. Watanabe, R. Horie, and N. Yamamoto. 2005. Hodgkin's lymphoma cells are efficiently engrafted and tumor marker CD30 is expressed with constitutive nuclear factor-kappaB activity in unconditioned NOD/SCID/γc mice. *Cancer Sci* 96:466-473.
19. Dewan, M. Z., M. Watanabe, K. Terashima, M. Aoki, T. Sata, M. Honda, M. Ito, S. Yamaoka, T. Watanabe, R. Horie, and N. Yamamoto. 2004. Prompt tumor formation and maintenance of constitutive NF-kappaB activity of multiple myeloma cells in NOD/SCID/γc^{null} mice. *Cancer Sci* 95:564-568.
20. Dick, J. E. 1996. Human stem cell assays in immune-deficient mice. *Curr Opin Hematol* 3:405-409.
21. Dick, J. E., C. Sirard, F. Pflumio, and T. Lapidot. 1992. Murine models of normal and neoplastic human haematopoiesis. *Cancer Surveys* 15:161-181.

22. DiSanto, J. P., W. Muller, D. Guy-Grand, A. Fischer, and K. Rajewsky. 1995. Lymphoid development in mice with a targeted deletion of the interleukin 2 receptor gamma chain. *Proc Natl Acad Sci USA* 92:377-381.
23. Doshi, M., M. Koyanagi, M. Nakahara, K. Saeki, K. Saeki, and A. Yuo. 2006. Identification of human neutrophils during experimentally induced inflammation in mice with transplanted CD34+ cells from human umbilical cord blood. *Int J Hematol* 84:231-237.
24. Fu, Y. X., H. Molina, M. Matsumoto, G. Huang, J. Min, and D. D. Chaplin. 1997. Lymphotoxin-alpha (LTα) supports development of splenic follicular structure that is required for IgG responses. *J Exp Med* 185:2111-2120.
25. Fugmann, S. D., A. I. Lee, P. E. Shockett, I. J. Villey, and D. G. Schatz. 2000. The RAG proteins and V(D)J recombination: complexes, ends, and transposition. *Annu Rev Immunol* 18:495-527.
26. Fujino, H., H. Hiramatsu, A. Tsuchiya, A. Niwa, H. Noma, M. Shiota, K. Umeda, M. Yoshimoto, M. Ito, T. Heike, and T. Nakahata. 2007. Human cord blood CD34+ cells develop into hepatocytes in the livers of NOD/SCID/γc^{null} mice through cell fusion. *FASEB J*. in press.
27. Gimeno, R., K. Weijer, A. Voordouw, C. H. Uittenbogaart, N. Legrand, N. L. Alves, E. Wijnands, B. Blom, and H. Spits. 2004. Monitoring the effect of gene silencing by RNA interference in human CD34+ cells injected into newborn RAG2$^{-/-}$ γc$^{-/-}$ mice: functional inactivation of p53 in developing T cells. *Blood* 104:3886-3893.
28. Giovanella, B. C., and J. Fogh. 1985. The nude mouse in cancer research. *Adv Cancer Res* 44:69-120.
29. Goldman, J. P., M. P. Blundell, L. Lopes, C. Kinnon, J. P. Di Santo, and A. J. Thrasher. 1998. Enhanced human cell engraftment in mice deficient in RAG2 and the common cytokine receptor gamma chain. *Br J Haematol* 103:335-342.
30. Gorantla, S., H. Sneller, L. Walters, J. G. Sharp, S. J. Pirruccello, J. T. West, C. Wood, S. Dewhurst, H. E. Gendelman, and L. Poluektova. 2007. Human immunodeficiency virus type 1 pathobiology studied in humanized BALB/c-Rag2$^{-/-}$ γc$^{-/-}$ mice. *J Virol* 81:2700-2712.
31. Greiner, D. L., R. A. Hesselton, and L. D. Shultz. 1998. SCID mouse models of human stem cell engraftment. *Stem Cells (Dayton, Ohio)* 16:166-177.
32. Habu, S., H. Fukui, K. Shimamura, M. Kasai, Y. Nagai, K. Okumura, and N. Tamaoki. 1981. In vivo effects of anti-asialo GM1. I. Reduction of NK activity and enhancement of transplanted tumor growth in nude mice. *J Immunol* 127:34-38.
33. Hesselton, R. M., D. L. Greiner, J. P. Mordes, T. V. Rajan, J. L. Sullivan, and L. D. Shultz. 1995. High levels of human peripheral blood mononuclear cell engraftment and enhanced susceptibility to human immunodeficiency virus type 1 infection in NOD/LtSz-scid/scid mice. *J Infect Dis* 172:974-982.
34. Heyman, R. A., E. Borrelli, J. Lesley, D. Anderson, D. D. Richman, S. M. Baird, R. Hyman, and R. M. Evans. 1989. Thymidine kinase obliteration: creation of transgenic mice with controlled immune deficiency. *Proc Natl Acad Sci USA* 86:2698-2702.
35. Hill, A. V. 1998. The immunogenetics of human infectious diseases. *Annu Rev Immunol* 16:593-617.
36. Hioki, K., T. Kuramochi, S. Endoh, E. Terada, Y. Ueyama, and M. Ito. 2001. Lack of B cell leakiness in BALB/cA-nu, scid double mutant mice. *Exp Anim* 50:67-72.
37. Hiramatsu, H., R. Nishikomori, T. Heike, M. Ito, K. Kobayashi, K. Katamura, and T. Nakahata. 2003. Complete reconstitution of human lymphocytes from cord blood CD34+ cells using the NOD/SCID/γc^{null} mice model. *Blood* 102:873-880.
38. Ikebe, M., K. Miyakawa, K. Takahashi, K. Ohbo, M. Nakamura, K. Sugamura, T. Suda, K. Yamamura, and K. Tomita. 1997. Lymphohaematopoietic abnormalities and systemic lymphoproliferative disorder in interleukin-2 receptor gamma chain-deficient mice. *Int J Exp Pathol* 78:133-148.
39. Isaacson, J. H., and B. M. Cattanach. 1962. Report. *Mouse News Letter* 27:31.
40. Ishikawa, F., M. Yasukawa, B. Lyons, S. Yoshida, T. Miyamoto, G. Yoshimoto, T. Watanabe, K. Akashi, L. D. Shultz, and M. Harada. 2005. Development of functional human blood and immune systems in NOD/SCID/IL2 receptor γ chain^{null} mice. *Blood* 106:1565-1573.

41. Ito, M., H. Hiramatsu, K. Kobayashi, K. Suzue, M. Kawahata, K. Hioki, Y. Ueyama, Y. Koyanagi, K. Sugamura, K. Tsuji, T. Heike, and T. Nakahata. 2002. NOD/SCID/γc^{null} mouse: an excellent recipient mouse model for engraftment of human cells. *Blood* 100:3175-3182.
42. Jacobs, H., P. Krimpenfort, M. Haks, J. Allen, B. Blom, C. Demolliere, A. Kruisbeek, H. Spits, and A. Berns. 1999. PIM1 reconstitutes thymus cellularity in interleukin 7- and common gamma chain-mutant mice and permits thymocyte maturation in Rag- but not CD3gamma-deficient mice. *J Exp Med* 190:1059-1068.
43. Jung, S., D. Unutmaz, P. Wong, G. Sano, K. De los Santos, T. Sparwasser, S. Wu, S. Vuthoori, K. Ko, F. Zavala, E. G. Pamer, D. R. Littman, and R. A. Lang. 2002. In vivo depletion of CD11c$^+$ dendritic cells abrogates priming of CD8$^+$ T cells by exogenous cell-associated antigens. *Immunity* 17:211-220.
44. Kagi, D., B. Ledermann, K. Burki, P. Seiler, B. Odermatt, K. J. Olsen, E. R. Podack, R. M. Zinkernagel, and H. Hengartner. 1994. Cytotoxicity mediated by T cells and natural killer cells is greatly impaired in perforin-deficient mice. *Nature* 369:31-37.
45. Kambe, N., H. Hiramatsu, M. Shimonaka, H. Fujino, R. Nishikomori, T. Heike, M. Ito, K. Kobayashi, Y. Ueyama, N. Matsuyoshi, Y. Miyachi, and T. Nakahata. 2004. Development of both human connective tissue-type and mucosal-type mast cells in mice from hematopoietic stem cells with identical distribution pattern to human body. *Blood* 103:860-867.
46. Kamel-Reid, S., and J. E. Dick. 1988. Engraftment of immune-deficient mice with human hematopoietic stem cells. *Science* 242:1706-1709.
47. Kamogawa, Y., L. A. Minasi, S. R. Carding, K. Bottomly, and R. A. Flavell. 1993. The relationship of IL-4- and IFN gamma-producing T cells studied by lineage ablation of IL-4-producing cells. *Cell* 75:985-995.
48. Kasai, M., M. Iwamori, Y. Nagai, K. Okumura, and T. Tada. 1980. A glycolipid on the surface of mouse natural killer cells. *Eur J Immunol* 10:175-180.
49. Kelly, E. M. 1957. Report. *Mouse News Letter* 16:36.
50. Kikutani, H., and S. Makino. 1992. The murine autoimmune diabetes model: NOD and related strains. *Adv Immunol* 51:285-322.
51. Kirberg, J., A. Berns, and H. von Boehmer. 1997. Peripheral T cell survival requires continual ligation of the T cell receptor to major histocompatibility complex-encoded molecules. *J Exp Med* 186:1269-1275.
52. Kirchgessner, C. U., C. K. Patil, J. W. Evans, C. A. Cuomo, L. M. Fried, T. Carter, M. A. Oettinger, and J. M. Brown. 1995. DNA-dependent kinase (p350) as a candidate gene for the murine SCID defect. *Science* 267:1178-1183.
53. Koyanagi, Y., Y. Tanaka, J. Kira, M. Ito, K. Hioki, N. Misawa, Y. Kawano, K. Yamasaki, R. Tanaka, Y. Suzuki, Y. Ueyama, E. Terada, T. Tanaka, M. Miyasaka, T. Kobayashi, Y. Kumazawa, and N. Yamamoto. 1997. Primary human immunodeficiency virus type 1 viremia and central nervous system invasion in a novel hu-PBL-immunodeficient mouse strain. *J Virol* 71:2417-2424.
54. Kuroiwa, Y., H. Yoshida, T. Ohshima, T. Shinohara, A. Ohguma, Y. Kazuki, M. Oshimura, I. Ishida, and K. Tomizuka. 2002. The use of chromosome-based vectors for animal transgenesis. *Gene Therapy* 9:708-712.
55. Lander, E. S., L. M. Linton, B. Birren, C. Nusbaum, M. C. Zody et al. 2001. Initial sequencing and analysis of the human genome. *Nature* 409:860-921.
56. Linder, C. C. 2006. Genetic variables that influence phenotype. *ILAR J* 47:132-140.
57. Lowry, P. A., L. D. Shultz, D. L. Greiner, R. M. Hesselton, E. L. Kittler, C. Y. Tiarks, S. S. Rao, J. Reilly, J. H. Leif, H. Ramshaw, F. M. Stewart, and P. J. Quesenberry. 1996. Improved engraftment of human cord blood stem cells in NOD/LtSz-scid/scid mice after irradiation or multiple-day injections into unirradiated recipients. *Biol Blood Marrow Transplant* 2:15-23.
58. Makino, S., K. Kunimoto, Y. Muraoka, Y. Mizushima, K. Katagiri, and Y. Tochino. 1980. Breeding of a non-obese, diabetic strain of mice. *Jikken Dobutsu* 29:1-13.
59. Markel, P., P. Shu, C. Ebeling, G. A. Carlson, D. L. Nagle, J. S. Smutko, and K. J. Moore. 1997. Theoretical and empirical issues for marker-assisted breeding of congenic mouse strains. *Nat Genet* 17:280-284.

60. Masuda, H., T. Maruyama, E. Hiratsu, J. Yamane, A. Iwanami, T. Nagashima, M. Ono, H. Miyoshi, H. J. Okano, M. Ito, N. Tamaoki, T. Nomura, H. Okano, Y. Matsuzaki, and Y. Yoshimura. 2007. Noninvasive and real-time assessment of reconstructed functional human endometrium in NOD/SCID/γc^{null} immunodeficient mice. *Proc Natl Acad Sci USA* 104:1925-1930.
61. Matsumura, T., Y. Kametani, K. Ando, Y. Hirano, I. Katano, R. Ito, M. Shiina, H. Tsukamoto, Y. Saito, Y. Tokuda, S. Kato, M. Ito, K. Motoyoshi, and S. Habu. 2003. Functional CD5+ B cells develop predominantly in the spleen of NOD/SCID/γc^{null} (NOG) mice transplanted either with human umbilical cord blood, bone marrow, or mobilized peripheral blood CD34+ cells. *Exp Hematol* 31:789-797.
62. Matsuura-Sawada, R., T. Murakami, Y. Ozawa, H. Nabeshima, J. Akahira, Y. Sato, Y. Koyanagi, M. Ito, Y. Terada, and K. Okamura. 2005. Reproduction of menstrual changes in transplanted human endometrial tissue in immunodeficient mice. *Hum Reprod* 20:1477-1484.
63. Mazurier, F., M. Doedens, O. I. Gan, and J. E. Dick. 2003. Characterization of cord blood hematopoietic stem cells. *Ann NY Acad Sci* 996:67-71.
64. McCune, J. M., R. Namikawa, H. Kaneshima, L. D. Shultz, M. Lieberman, and I. L. Weissman. 1988. The SCID-hu mouse: murine model for the analysis of human hematolymphoid differentiation and function. *Science* 241:1632-1639.
65. Melkus, M. W., J. D. Estes, A. Padgett-Thomas, J. Gatlin, P. W. Denton, F. A. Othieno, A. K. Wege, A. T. Haase, and J. V. Garcia. 2006. Humanized mice mount specific adaptive and innate immune responses to EBV and TSST-1. *Nat Med* 12:1316-1322.
66. Meuleman, P., L. Libbrecht, R. De Vos, B. de Hemptinne, K. Gevaert, J. Vandekerckhove, T. Roskams, and G. Leroux-Roels. 2005. Morphological and biochemical characterization of a human liver in a uPA-SCID mouse chimera. *Hepatology* 41:847-856.
67. Miller, R. D., J. Hogg, J. H. Ozaki, D. Gell, S. P. Jackson, and R. Riblet. 1995. Gene for the catalytic subunit of mouse DNA-dependent protein kinase maps to the scid locus. *Proc Natl Acad Sci USA* 92:10792-10795.
68. Miyakawa, Y., Fukuchi, Y., Ito, M., Kobayashi, K., Kuramochi, T., Ikeda, Y., Takebe, Y., Tanaka, T., Miyasaka, M., Tanaka, N., Tamaoki, N., Nomura, T., Ueyama, Y., Shimamura, K. 1996. Establishment of human granulocyte-macrophage colony stimulating factor produing transgenic SCID mice. *Br J Haematol* 95:437-442.
69. Miyakawa, Y., Y. Ohnishi, M. Tomisawa, M. Monnai, K. Kohmura, Y. Ueyama, M. Ito, Y. Ikeda, M. Kizaki, and M. Nakamura. 2004. Establishment of a new model of human multiple myeloma using NOD/SCID/γc^{null} (NOG) mice. *Biochem Biophys Res Commun* 313:258-262.
70. Mombaerts, P., J. Iacomini, R. S. Johnson, K. Herrup, S. Tonegawa, and V. E. Papaioannou. 1992. RAG-1-deficient mice have no mature B and T lymphocytes. *Cell* 68:869-877.
71. Mosier, D. E., R. J. Gulizia, S. M. Baird, S. Spector, D. Spector, T. J. Kipps, R. I. Fox, D. A. Carson, N. Cooper, D. D. Richman et al. 1989. Studies of HIV infection and the development of Epstein-Barr virus-related B cell lymphomas following transfer of human lymphocytes to mice with severe combined immunodeficiency. *Curr Top Microbiol Immunol* 152:195-199.
72. Mosier, D. E., R. J. Gulizia, S. M. Baird, and D. B. Wilson. 1988. Transfer of a functional human immune system to mice with severe combined immunodeficiency. *Nature* 335:256-259.
73. Muguruma, Y., T. Yahata, H. Miyatake, T. Sato, T. Uno, J. Itoh, S. Kato, M. Ito, T. Hotta, and K. Ando. 2006. Reconstitution of the functional human hematopoietic microenvironment derived from human mesenchymal stem cells in the murine bone marrow compartment. *Blood* 107:1878-1887.
74. Mule, J. J., D. L. Jicha, P. M. Aebersold, W. D. Travis, and S. A. Rosenberg. 1991. Disseminated human malignant melanoma in congenitally immune-deficient (bg/nu/xid) mice. *J NCI* 83:350-355.
75. Nakamura, T., Y. Miyakawa, A. Miyamura, A. Yamane, H. Suzuki, M. Ito, Y. Ohnishi, N. Ishiwata, Y. Ikeda, and N. Tsuruzoe. 2006. A novel nonpeptidyl human c-Mpl activator stimulates human megakaryopoiesis and thrombopoiesis. *Blood* 107:4300-4307.

76. Nakata, H., K. Maeda, T. Miyakawa, S. Shibayama, M. Matsuo, Y. Takaoka, M. Ito, Y. Koyanagi, and H. Mitsuya. 2005. Potent anti-R5 human immunodeficiency virus type 1 effects of a CCR5 antagonist, AK602/ONO4128/GW873140, in a novel human peripheral blood mononuclear cell nonobese diabetic-SCID, interleukin-2 receptor gamma-chain-knocked-out AIDS mouse model. *J Virol* 79:2087-2096.
77. Ninomiya, M., A. Abe, T. Yokozawa, K. Ozeki, K. Yamamoto, M. Ito, H. Kiyoi, N. Emi, and T. Naoe. 2006. Establishment of a myeloid leukemia cell line, TRL-01, with MLL-ENL fusion gene. *Cancer Genet Cytogenet* 169:1-11.
78. Ohbo, K., T. Suda, M. Hashiyama, A. Mantani, M. Ikebe, K. Miyakawa, M. Moriyama, M. Nakamura, M. Katsuki, K. Takahashi, K. Yamamura, and K. Sugamura. 1996. Modulation of hematopoiesis in mice with a truncated mutant of the interleukin-2 receptor gamma chain. *Blood* 87:956-967.
79. Ohteki, T., Fukao, T., Suzue, K., Maki, C., Ito, M., Nakamura, M. and Koyasu, S. 1999. Interleukin-12 dependent interferon-γ production by CD8a$^+$ lymphoid dendritic cells. *J Exp Med* 189:1981-1986.
80. Passegue, E., C. H. Jamieson, L. E. Ailles, and I. L. Weissman. 2003. Normal and leukemic hematopoiesis: are leukemias a stem cell disorder or a reacquisition of stem cell characteristics? *Proc Natl Acad Sci USA* 100 Suppl 1:11842-11849.
81. Pearson, T., D. L. Greiner, and L. D. Shultz. 2007. Humanized SCID mouse models for biomedical research. This volume.
82. Pflumio, F., B. Izac, A. Katz, L. D. Shultz, W. Vainchenker, and L. Coulombel. 1996. Phenotype and function of human hematopoietic cells engrafting immune-deficient CB17-severe combined immunodeficiency mice and nonobese diabetic-severe combined immuno-deficiency mice after transplantation of human cord blood mononuclear cells. *Blood* 88:3731-3740.
83. Prochazka, M., H. R. Gaskins, L. D. Shultz, and E. H. Leiter. 1992. The nonobese diabetic scid mouse: model for spontaneous thymomagenesis associated with immunodeficiency. *Proc Natl Acad Sci USA* 89:3290-3294.
84. Rao, M. K., and M. F. Wilkinson. 2006. Tissue-specific and cell type-specific RNA interference in vivo. *Nat Proto* 1:1494-1501.
85. Saito, M., T. Iwaki, C. Taya, H. Yonekawa, M. Noda, Y. Inui, E. Mekada, Y. Kimata, A. Tsuru, and K. Kohno. 2001. Diphtheria toxin receptor-mediated conditional and targeted cell ablation in transgenic mice. *Nat Biotechnol* 19:746-750.
86. Seibler, J., A. Kleinridders, B. Kuter-Luks, S. Niehaves, J. C. Bruning, and F. Schwenk. 2007. Reversible gene knockdown in mice using a tight, inducible shRNA expression system. *Nucleic Acids Res* 35:e54.
87. Seita, J., H. Ema, J. Ooehara, S. Yamazaki, Y. Tadokoro, A. Yamasaki, K. Eto, S. Takaki, K. Takatsu, and H. Nakauchi. 2007. Lnk negatively regulates self-renewal of hematopoietic stem cells by modifying thrombopoietin-mediated signal transduction. *Proc Natl Acad Sci USA* 104:2349-2354.
88. Shinkai, Y., G. Rathbun, K. P. Lam, E. M. Oltz, V. Stewart, M. Mendelsohn, J. Charron, M. Datta, F. Young, A. M. Stall, et al. 1992. RAG-2-deficient mice lack mature lymphocytes owing to inability to initiate V(D)J rearrangement. *Cell* 68:855-867.
89. Shresta, S., D. M. MacIvor, J. W. Heusel, J. H. Russell, and T. J. Ley. 1995. Natural killer and lymphokine-activated killer cells require granzyme B for the rapid induction of apoptosis in susceptible target cells. *Proc Natl Acad Sci USA* 92:5679-5683.
90. Shultz, L. D., S. Banuelos, B. Lyons, R. Samuels, L. Burzenski, B. Gott, P. Lang, J. Leif, M. Appel, A. Rossini, and D. L. Greiner. 2003. NOD/LtSz-Rag1^{null}Pfp^{null} mice: a new model system with increased levels of human peripheral leukocyte and hematopoietic stem-cell engraftment. *Transplantation* 76:1036-1042.
91. Shultz, L. D., F. Ishikawa, and D. L. Greiner. 2007. Humanized mice in translational biomedical research. *Nat Rev* 7:118-130.
92. Shultz, L. D., B. L. Lyons, L. M. Burzenski, B. Gott, X. Chen, S. Chaleff, M. Kotb, S. D. Gillies, M. King, J. Mangada, D. L. Greiner, and R. Handgretinger. 2005. Human lymphoid

and myeloid cell development in NOD/LtSz-scid IL2R gamma null mice engrafted with mobilized human hemopoietic stem cells. *J Immunol* 174:6477-6489.
93. Shultz, L. D., P. A. Schweitzer, S. W. Christianson, B. Gott, I. B. Schweitzer, B. Tennent, S. McKenna, L. Mobraaten, T. V. Rajan, D. L. Greiner et al. 1995. Multiple defects in innate and adaptive immunologic function in NOD/LtSz-scid mice. *J Immunol* 154:180-191.
94. Steinsvik, T. E., P. I. Gaarder, I. S. Aaberge, and Lovik. 1995. Engraftment and humoral immunity in SCID and RAG-2-deficient mice transplanted with human peripheral blood lymphocytes. *Scand J Immunol* 42:607-616.
95. Suematsu, S., and T. Watanabe. 2004. Generation of a synthetic lymphoid tissue-like organoid in mice. *Nat Biotechnol* 22:1539-1545.
96. Suemizu, H., M. Monnai, Y. Ohnishi, M. Ito, N. Tamaoki, and M. Nakamaura. 2007. Identification of a key molecular regulator of liver metastasis in human pancreatic carcinoma using a novel quantitative model of metastasis in NOD/SCID/γc^{null} (NOG) mice. *Int J Oncol* 31:741-751..
97. Sugamura, K., H. Asao, M. Kondo, N. Tanaka, N. Ishii, K. Ohbo, M. Nakamura, and T. Takeshita. 1996. The interleukin-2 receptor gamma chain: its role in the multiple cytokine receptor complexes and T cell development in XSCID. *Annu Rev Immunol* 14:179-205.
98. Suwa, H., T. Tanaka, F. Kitamura, T. Shiohara, K. Kuida, and M. Miyasaka. 1995. Dysregulated expression of the IL-2 receptor beta-chain abrogates development of NK cells and Thy-1+ dendritic epidermal cells in transgenic mice. *Int Immunol* 7:1441-1449.
99. Taieb, J., N. Chaput, C. Menard, L. Apetoh, E. Ullrich, M. Bonmort, M. Pequignot, N. Casares, M. Terme, C. Flament, P. Opolon, Y. Lecluse, D. Metivier, E. Tomasello, E. Vivier, F. Ghiringhelli, F. Martin, D. Klatzmann, T. Poynard, T. Tursz, G. Raposo, H. Yagita, B. Ryffel, G. Kroemer, and L. Zitvogel. 2006. A novel dendritic cell subset involved in tumor immunosurveillance. *Nat Med* 12:214-219.
100. Tanaka, T., F. Kitamura, Y. Nagasaka, K. Kuida, H. Suwa, and M. Miyasaka. 1993. Selective long-term elimination of natural killer cells in vivo by an anti-interleukin 2 receptor beta chain monoclonal antibody in mice. *J Exp Med* 178:1103-1107.
101. Tateno, C., Y. Yoshizane, N. Saito, M. Kataoka, R. Utoh, C. Yamasaki, A. Tachibana, Y. Soeno, K. Asahina, H. Hino, T. Asahara, T. Yokoi, T. Furukawa, and K. Yoshizato. 2004. Near completely humanized liver in mice shows human-type metabolic responses to drugs. *Am J Pathol* 165:901-912.
102. Torbett, B. E., G. Picchio, and D. E. Mosier. 1991. hu-PBL-SCID mice: a model for human immune function, AIDS, and lymphomagenesis. *Immunol Rev* 124:139-164.
103. Traggiai, E., L. Chicha, L. Mazzucchelli, L. Bronz, J. C. Piffaretti, A. Lanzavecchia, and M. G. Manz. 2004. Development of a human adaptive immune system in cord blood cell-transplanted mice. *Science* 304:104-107.
104. Ueda, T., H. Yoshino, K. Kobayashi, M. Kawahata, Y. Ebihara, M. Ito, S. Asano, T. Nakahata, and K. Tsuji. 2000. Hematopoietic repopulating ability of cord blood CD34+ cells in NOD/Shi-scid mice. *Stem Cells (Dayton, Ohio)* 18:204-213.
105. Warner, J. K., J. C. Wang, K. J. Hope, L. Jin, and J. E. Dick. 2004. Concepts of human leukemic development. *Oncogene* 23:7164-7177.
106. Watanabe, S., K. Terashima, S. Ohta, S. Horibata, M. Yajima, Y. Shiozawa, M. Z. Dewan, Z. Yu, M. Ito, T. Morio, N. Shimizu, M. Honda, and N. Yamamoto. 2007. Hematopoietic stem cell-engrafted NOD/SCID/IL2Rgamma null mice develop human lymphoid systems and induce long-lasting HIV-1 infection with specific humoral immune responses. *Blood* 109:212-218.
107. Waterston, R. H., K. Lindblad-Toh, E. Birney, J. Rogers, J. F. Abril, P. Agarwal, R. Agarwala et al. 2002. Initial sequencing and comparative analysis of the mouse genome. *Nature* 420:520-562.
108. Widney, D. P., Y. R. Xia, A. J. Lusis, and J. B. Smith. 2000. The murine chemokine CXCL11 (IFN-inducible T cell alpha chemoattractant) is an IFN-gamma- and lipopolysaccharide-inducible glucocorticoid-attenuated response gene expressed in lung and other tissues during endotoxemia. *J Immunol* 164:6322-6331.

109. Wills-Karp, M. 1999. Immunologic basis of antigen-induced airway hyperresponsiveness. *Annu Rev Immunol* 17:255-281.
110. Wong, G. T. 2002. Speed congenics: applications for transgenic and knock-out mouse strains. *Neuropeptides* 36:230-236.
111. Yahata, T., K. Ando, H. Miyatake, T. Uno, T. Sato, M. Ito, S. Kato, and T. Hotta. 2004. Competitive repopulation assay of two gene-marked cord blood units in NOD/SCID/gammac^{null} mice. *Mol Ther* 10:882-891.
112. Yahata, T., K. Ando, Y. Nakamura, Y. Ueyama, K. Shimamura, N. Tamaoki, S. Kato, and T. Hotta. 2002. Functional human T lymphocyte development from cord blood CD34+ cells in nonobese diabetic/Shi-scid, IL-2 receptor gamma null mice. *J Immunol* 169:204-209.
113. Yahata, T., S. Yumino, Y. Seng, H. Miyatake, T. Uno, Y. Muguruma, M. Ito, H. Miyoshi, S. Kato, T. Hotta, and K. Ando. 2006. Clonal analysis of thymus-repopulating cells presents direct evidence for self-renewal division of human hematopoietic stem cells. *Blood* 108:2446-2454.
114. Yamazaki, S., A. Iwama, Y. Morita, K. Eto, H. Ema, and H. Nakauchi. 2007. Cytokine signaling, lipid raft clustering, and HSC hibernation. *Ann NY Acad Sci* 1106:54-63.
115. Yoshino, H., T. Ueda, M. Kawahata, K. Kobayashi, Y. Ebihara, A. Manabe, R. Tanaka, M. Ito, S. Asano, T. Nakahata, and K. Tsuji. 2000. Natural killer cell depletion by anti-asialo GM1 antiserum treatment enhances human hematopoietic stem cell engraftment in NOD/Shi-scid mice. *Bone Marrow Transplant* 26:1211-1216.
116. Zijlstra, M., E. Li, F. Sajjadi, S. Subramani, and R. Jaenisch. 1989. Germ-line transmission of a disrupted beta 2-microglobulin gene produced by homologous recombination in embryonic stem cells. *Nature* 342:435-438.

Humanizing Bone Marrow in Immune-Deficient Mice

K. Ando(✉), Y. Muguruma, and T. Yahata

1	Introduction	78
2	Clonal Analysis of Human Hematopoietic Stem Cells in NOG Mice	79
3	Humanizing Hematopoietic Microenvironment in Mice	81
4	Conclusion	83
References		84

Abstract Humanized mice are useful for studying human hematopoietic stem cells (HSCs) and their niche. In particular, clonal study of human HSC enables precise comparison of in vivo behavior between murine and human HSCs. A single HSC is able to reconstitute hematopoiesis even after serial transplantations in mice. While the life span of somatic cells is over that of individual in mice, this is not the case in humans. Clonal studies of human HSCs clearly demonstrated their aging in hosts. Since murine studies have demonstrated that HSCs are protected from aging by their niche in bone marrow, the humanizing niche model will reveal the precise mechanism by which human HSCs are protected from exhaustion in vivo. Direct transplantation of human mesenchymal stem cells into mouse bone marrow results in reconstitution of the functional human hematopoietic microenvironment comprised of pericytes, myofibroblasts, reticular cells, osteocytes in bone, bone-lining osteoblasts, and endothelial cells. These humanized mouse models are essential for testing whether the insights on hematopoiesis from mouse studies are applicable to humans before clinical application.

Abbreviations ALP: alkaline phosphatase; BM: bone marrow; DP: double positive; FACS: fluorescence-activated cell sorter; GFP: green fluorescent protein; HME: Hematopoietic microenvironment; HMRC: hematopoietic microenvironment-reconstituting cells; HSC: hematopoietic stem cell; IBMT:

K. Ando
Division of Hematopoiesis, Department of Hematology, Research Center of Regenerative Medicine, Tokai University School of Medicine, Bohseidai, Isehara, Kanagawa 259-1193, Japan
andok@keyaki.cc.u-tokai.ac.jp

intra-bone marrow transplantation; IL-2: interleukin 2; LAM-PCR: linear amplification-mediated-PCR; MSC: mesenchymal stem cell; PCR: polymerase chain reaction; PHA: phytohemagglutinin; SDF-1: stromal cell-derived factor 1; SRC: severe combined immunodeficiency (*scid*) mouse-repopulating cells; YFP: yellow fluorescent protein

1 Introduction

Production of blood cells over a lifetime is sustained by hematopoietic stem cells (HSCs) that are defined as cells having both the capacity for self-renewal and the ability to produce all mature hematopoietic lineages. The presence of HSCs was first demonstrated retrospectively by a gene-marking study [28]. Prospective identification was impossible until the successful purification of a single HSC with defined surface markers [30, 39, 50]. Then clonal studies of serial transplanted HSCs demonstrated that they intrinsically limit their potential for self-renewal, and the mean activity of individual stem cells was reduced with cell division [15]. Therefore, HSCs cannot maintain their immaturity by themselves but require a special environment, or niche, to keep themselves dormant in vivo for protection from exhaustion. Recent studies identify two types of stem cell niche in bone marrow (BM), the osteoblastic niche and the vascular niche, and several key molecules such as jagged-1, N-cadherin, CXCR-4, and angiopoietin-1 for niche function [3, 9, 25, 48, 55]. All of these essential findings on HSCs and the niche come from mouse studies. It is not clear whether these findings can be extrapolated to large animals including humans, since it is reported that HSCs behave differently between mice and felines [1]. Therefore, these findings must be tested in human HSCs before translation into clinical application.

To evaluate human HSC activity, xenotransplantation models are recognized as the most reliable surrogate assay. Immune-deficient mice have been widely used as recipients, since myeloid and lymphoid reconstitution can be easily attained in these mice by the transplantation of human HSCs [27, 45]. Severe combined immunodeficiency (*scid*) mouse-repopulating cell (SRC) assay has demonstrated characteristic features of human HSCs such as self-renewal and multilineage differentiation ability, frequency, surface markers, organization of hierarchy, and dynamic behaviors in vivo [5, 6, 7, 12, 16, 20, 31, 40, 49]. On the other hand, the human hematopoietic microenvironment has not been studied in vivo because of the lack of an appropriate model. Recently, Muguruma et al. established a human mesenchymal stem cell (MSC)-derived hematopoietic microenvironment-reconstituting cell (HMRC) model in immune-deficient mice [35].

In this review, we summarize recent results on clonal analysis of human HSCs and human microenvironment in the xenotransplantation model.

2 Clonal Analysis of Human Hematopoietic Stem Cells in NOG Mice

Unlike murine HSCs that have been purified and analyzed at the single-cell level, retrovirus-mediated gene marking is the only strategy for in vivo clonal analysis of human HSCs. Since retrovirus vectors essentially integrate at random, each genomic integration site serves as a distinct clonal marker that can be used to trace the progeny of individual stem cells after transplantation. Gene-marking studies in SRC assays have successfully elucidated the in vivo kinetics of human HSCs such as proliferation, self-renewal, and multilineage differentiation [2, 18, 32, 33, 37].

A major shortcoming of using the NOD/SCID mouse model is a lack of reproducible human T-lymphocyte repopulation. Consequently, the multilineage differentiation capacity of SRCs in NOD/SCID recipients has been assessed by reconstitution of only B-lymphoid and myeloid lineages. Since a close relationship between B-lymphocyte and macrophage differentiation has been indicated [8, 14, 21, 34], current analyses cannot clearly distinguish true HSCs from lineage-restricted progenitors such as B-lymphocyte/macrophage progenitors. Thus, the multilineage differentiation and self-renewal of HSCs represented by a single SRC are yet to be proven.

NOD/SCID/γc^{null} (NOG) mice were recently generated by crossing NOD/Shi-scid mice with mice expressing a form of the IL-2R γ chain lacking the cytoplasmic region that were reported to have defective NK cells [22]. They have defective T, B, and NK cell activities so that they demonstrate human T cell development in their thymus, in addition to myeloid and B-lymphoid reconstitution, when transplanted with $CD34^+$ cells. Furthermore, these T cells bear polyclonal $\alpha\beta$ TCR and respond not only to mitogenic stimuli, such as PHA and IL-2, but also to allogenic human cells. These results indicate that functional human T lymphocytes can be reconstituted from $CD34^+$ cells in NOG mice [19, 46, 52].

HSCs can be identified as thymus-repopulating cells and distinguished from short-lived oligo- or monopotent progenitors in NOG mice. Thymopoiesis requires constant recruitment of progenitors into the thymus, which eventually produces mature T-lymphocytes in a relatively short period of time [17]. Therefore, to maintain thymopoiesis in recipient mice, transplanted HSCs must divide without loss of thymus-repopulating activity. While several classes of SRCs that differ in their proliferative and self-renewal potential have been reported [16, 20, 31], analyzing the thymus-repopulating activity of these cells provides a unique way to distinguish and identify long-term self-renewing stem cells within the SRCs.

Yahata et al. established a novel strategy to analyze both self-renewal and multilineage differentiation of a single human thymus-repopulating SRC clone in NOG recipient mice [54] using linear amplification-mediated-PCR (LAM-PCR) that verifies individual genomic virus integration sites by direct sequencing [44] (Fig. 1). The identification of specific clones in fluorescence-activated cell sorter (FACS)-sorted lymphomyeloid lineage populations by their unique molecular markers allowed us to assess how individual clones contribute to the specific lineages during long-term

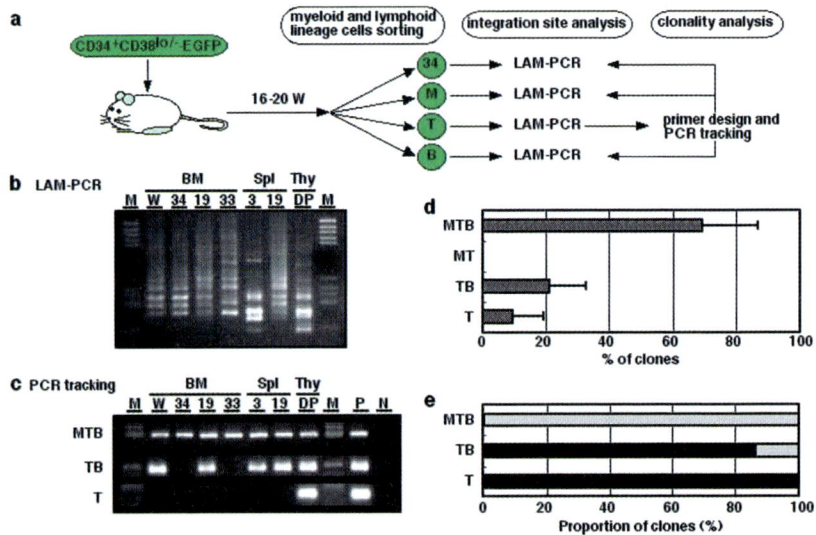

Fig. 1 Clonal analysis of primary transplanted SRCs [54]. **a** Study design. **b** Representative LAM-PCR profiles of SRCs. Each band represents a different insertion locus in the assayed material. **c** DP-derived T-lymphoid insertion sites were traced by PCR. The clones detected in all lymphomyeloid lineage cells were designated as multipotent type (MTB). Clones restricted in T-lymphoid and B-lymphoid cells were p-TB. Clones detected in T-lymphoid cells were p-T. **d** Relative frequencies of each clone type detected in primary SRCs. **e** The proportion of clones detected in the CD34+ cell population. *Gray bars* represent the clones detected in CD34+ cells. *Black bars* represent the clones not detected in CD34+ cells

hematopoiesis in vivo. They focused on the CD4/CD8 double-positive (DP) immature thymocyte population as a starting point for analysis of the clonal capacity of human HSCs. The study presents direct clonal evidence that a single human HSC had the capacity to produce lymphoid and myeloid lineage cells and that self-renewal division of multilineage clones resulted in expansion of SRC populations in primary and secondary NOG recipients during long-term hematopoiesis.

The finding that the same multipotent HSC clone was detected in paired secondary recipients (MTB-MTB type) indicates that a single SRC clone can self-replicate to produce two daughter cells with multilineage differentiation and self-renewal potential, leading to the in vivo expansion of SRCs. On the other hand, one of the daughter clones of the pair was no longer found in the CD34+ cell population in approximately half of MTB-MTB clone pairs. Furthermore, the phenotype was not retained in 11.1% of MTB-MTB clones. Considering that 100% of MTB clones in primary recipients possess the stem cell phenotype, these results indicate that SRCs with the stem cell phenotype progressively decrease during serial transplantation, leading to exhaustion of SRCs. This is consistent with the finding that the proportion of clones with the stem cell phenotype decreased as the clone committed to specific lineages. By assessing the phenotype of self-replicated multilineage clone pairs, determined by the presence of common clones in the CD34+ cell population, they were able to reveal the status of HSCs during aging. It is possible that the extensive replication required for hematopoietic reconstitution in the recipient may result in loss of the stem cell phenotype. The data indicate that although the total SRC population appears to expand, the ability of individual SRCs may become restricted during long-term hematopoiesis in vivo.

3   Humanizing Hematopoietic Microenvironment in Mice

The stem cell niche is a key determinant of stem cell development. We are beginning to understand the murine HSC niche and the molecular mechanisms that govern the fate of murine HSCs [3, 9, 25, 48, 55], but there exists a paucity of data on the cellular and molecular microenvironmental regulation of human hematopoiesis in vivo , due largely to a lack of good experimental tools. Although the identification of SRCs has facilitated detailed characterization of human HSCs in vivo, the key niches that function in human cell repopulation have not been identified.

Mesenchymal stem cells (MSCs) present in BM are thought to give rise to cells that constitute the hematopoietic microenvironment (HME) [43]. MSCs have been isolated from BM and various tissues from humans and many other species, expanded in culture, and shown to differentiate into osteocytes, chondrocytes, adipocytes, and myoblasts under defined conditions in vitro [41]. In culture, MSCs produce a number of cytokines and extracellular matrix proteins and express cell adhesion molecules, all of which are involved in the regulation of hematopoiesis [13, 29]. They also support the development of hematopoietic colonies in vitro. However, in contrast to HSCs, MSCs have only been defined and isolated by physical and functional properties in vitro. Consequently, little is known about their phenotypic and functional characteristics in vivo.

Systemic administration of MSCs for facilitation of BM transplantation has been proposed based on the in vitro characteristics of MSCs [23]. In recent studies, cotransplantation of human MSCs and HSCs resulted in increased chimerism or accelerated hematopoietic recovery (or both) in animal models and in humans [26, 38], suggesting a role for MSCs in the engraftment and repopulation of HSCs. Although the existence of donor MSCs has been documented in the BM of recipient animals after MSC infusion [4], the methods used to detect engraftment, such as polymerase chain reaction (PCR) or staining of cytospin samples, could not unambiguously distinguish engraftment from cell survival or nonspecific lodgment on the vascular bed. There is no physical evidence that transplanted human MSCs have indeed engrafted in the BM of adult animals and directly participated in the enhanced engraftment of HSCs.

To assess the engraftment, spatial distribution, and lineage commitment of MSCs as well as their roles in hematopoiesis in vivo, Muguruma et al. transplanted green fluorescent protein (GFP)-marked human MSCs into the tibiae of NOD/SCID mice by intra-bone marrow transplantation (IBMT) [35], a method previously

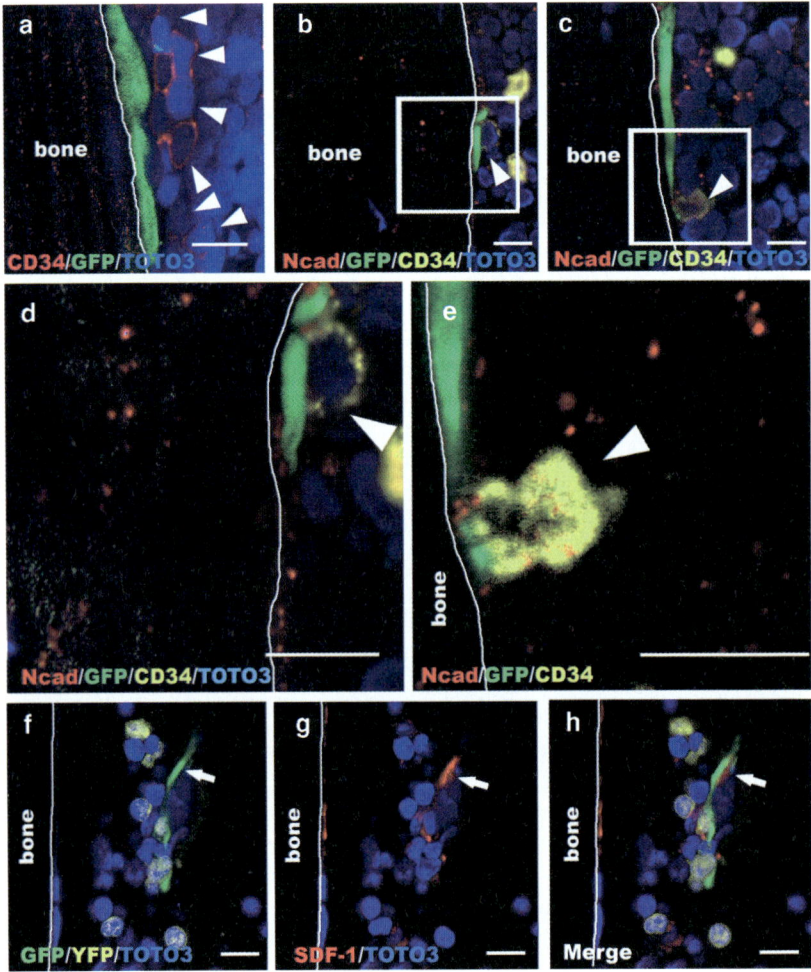

Fig. 2 Expression of N-cadherin and SDF-1 by HMRCs that interact with human hematopoietic cells [35]. **a** $CD34^+$ cells (*arrowheads*) appear to colonize near the bone-lining HMRCs. **b–c** Bone-lining HMRCs colocalize with $CD34^+$ cells (*arrowheads*) through the asymmetrical expression of N-cadherin. **d–e** Higher magnifications of **b** and **c**. **f–h** An HMRC in the endosteal hematopoietic parenchyma (*arrow*) expresses SDF-1 and interacts with a few YFP^+ hematopoietic cells. All *bars* represent 10 μm

shown to improve the engraftment of both hematopoietic and nonhematopoietic cells in mice [24, 53].

The phenotypes of transplanted MSCs and their progeny at 10 weeks after transplantation were investigated in detail. MSCs were preferentially localized to the endosteal region (73.8 ± 18.4%), frequently within five cells from the surface of the bone. When human MSC-derived GFP-expressing cells (GFP^+ cells) were found away from the endosteum, they were often associated with the vasculature.

Those vasculature-associated GFP$^+$ cells expressed α-SM actin, the expression of which has been documented in pericytes, SM cells of the vascular wall, as well as myofibroblasts in BM [11]. A total of 59.9 ± 21.6% of GFP$^+$ cells found in BM were positive for α-SM actin.

Two other types of GFP$^+$ cells that were negative for α-SM actin were also present in BM: flattened cells located in the hematopoietic cords but not specifically associated with the vasculature and cells characterized by long cytoplasmic extensions, so-called reticular cells that are considered to be the predominant cells of the HME [51]. BM was often interspersed with a fine network of cell processes that expressed alkaline phosphatase (ALP), an enzyme that distinguishes reticular cells from the stromal component of acid phosphatase- or nonspecific esterase-expressing macrophages. A total of 28.2 ± 11.2% of GFP$^+$ cells in BM were ALP positive.

In addition, GFP$^+$ cells were found within or on the surface of the bone. Cells in the bone stained positive for osteocalcin, a specific marker of mature osteoblasts and osteocytes, indicating an active participation in skeletal remodeling. Furthermore, GFP$^+$ cells on the bone surface resembled spindle-shaped osteoblasts, a key component of the stem cell niche in murine hematopoiesis [3, 9, 55]. These cells expressed osteopontin and N-cadherin, both of which are involved in regulating murine HSCs [36, 47]. These results indicate that within 10 weeks after transplantation, human MSCs differentiated into pericytes, myofibroblasts, reticular cells, osteocytes in bone, bone-lining osteoblasts, and endothelial cells, which constitute the three-dimensional structure of hematopoietic parenchyma and provide the milieu of hematopoiesis. These cells were designated as human MSC-derived hematopoietic microenvironment-reconstituting cells (HMRCs).

CD34$^+$ cells adhered to HMRCs on the bone surface and appeared to proliferate along the endosteal surface, suggesting the existence of specific local signals between CD34$^+$ cells and bone-lining HMRCs (Fig. 2a). Bone-lining HMRCs associated with CD34$^+$ cells through the colocalization of N-cadherin (Fig. 2b-e). In addition, an HMRC in the endosteal hematopoietic parenchyma expressed stromal cell-derived factor 1 (SDF-1) and interacted with a few YFP$^+$ human hematopoietic cells (Fig. 2f-h), although SDF-1 was not detected in ex vivo expanded MSCs by immunofluorescence analysis. In BM, SDF-1 is constitutively expressed by osteoblasts, endothelial cells, and BM stromal cells [42]. In addition to its well-established role in homing and retention of HSCs in BM, SDF-1 has been implicated in regulating the status of primitive HSCs both in vitro and in vivo [10]. Therefore, HMRCs may contribute to the maintenance of primitive human HSCs through N-cadherin-mediated interactions and the production of SDF-1.

4 Conclusion

Clonal study of human HSCs enables precise comparison of in vivo behavior between murine and human HSCs. A single HSC is able to reconstitute hematopoiesis even after serial transplantations in mice. While the life span of somatic cells is over that of individual in mice, this is not the case in humans. Clonal studies of human HSCs clearly demonstrated their aging in hosts. Since mouse studies have been demonstrated that HSCs are protected from aging by their niche in BM, the humanizing niche model will reveal the precise mechanism by which human HSCs are protected from exhaustion in vivo.

References

1. Abkowitz JL, Golinelli D, Harrison DE, Guttorp P. In vivo kinetics of murine hematopoietic stem cells. Blood 2000; 96:3399-3405
2. Ailles L, Schmidt M, Santoni de Sio FR, Glimm H, Cavalieri S, Bruno S, Piacibello W, Von Kalle C, Naldini L. Molecular evidence of lentiviral vector-mediated gene transfer into human self-renewing, multi-potent, long-term NOD/SCID repopulating hematopoietic cells. Mol Ther 2002; 6:615-626
3. Arai F, Hirao A, Ohmura M et al. Tie2/angiopoietin-1 signaling regulates hematopoietic stem cell quiescence in the bone marrow niche. Cell 2004; 118:149-161
4. Bensidhoum M, Chapel A, Francois S et al. Homing of in vitro expanded Stro-1− or Stro-1+ human mesenchymal stem cells into the NOD/SCID mouse and their role in supporting human CD34 cell engraftment. Blood 2004;103:3313-3319
5. Bhatia M, Wang JC, Kapp U, Bonnet D, Dick JE. Purification of primitive human hematopoietic cells capable of repopulating immune-deficient mice. Proc Natl Acad Sci USA 1997; 94:5320-5325
6. Bhatia M et al. Quantitative analysis reveals expansion of human hematopoietic repopulating cells after short-term ex vivo culture. J Exp Med 1997; 186:619-624
7. Bhatia M et al. A newly discovered class of human hematopoietic cells with SCID-repopulating activity. Nat Med 1998; 4:1038-1045
8. Borrello MA, Phipps RP. The B/macrophage cell: an elusive link between CD5+ B lymphocytes and macrophages. Immunol Today 1996; 17:471-475
9. Calvi LM, Adams GB, Weibrecht KW et al. Osteoblastic cells regulate the haematopoietic stem cell niche. Nature 2003; 425:841-846
10. Cashman J, Clark-Lewis I, Eaves A, Eaves C. Stromal-derived factor 1 inhibits the cycling of very primitive human hematopoietic cells in vitro and in NOD/SCID mice. Blood 2002; 99:792-799
11. Charbord P, Tavian M, Humeau L, Peault B. Early ontogeny of the human marrow from long bones: an immunohistochemical study of hematopoiesis and its microenvironment. Blood 1996; 87:4109-4119
12. Christopherson KW, Hangoc G, Mantel CR, Broxmeyer HE. Modulation of hematopoietic stem cell homing and engraftment by CD26. Science 2004; 305:1000-1003
13. Conget PA, Minguell JJ. Phenotypical and functional properties of human bone marrow mesenchymal progenitor cells. J Cell Physiol 1999; 181:67-73
14. Cumano A, Paige CJ, Iscove NN, Brady G. Bipotential precursors of B cells and macrophages in murine fetal liver. Nature 1992; 356:612-615
15. Ema H, Sudo K, Seita J, Matsubara A, Morita Y, Osawa M, Takatsu K, Takaki S, Nakauchi H. Quantification of self-renewal capacity in single hematopoietic stem cells from normal and lnk-deficient mice. Dev Cell 2005; 8:907-914
16. Glimm H, Eisterer W, Lee K, Cashman J, Holyoake TL, Nicolini F, Shultz LD, von Kalle C, Eaves CJ. Previously undetected human hematopoietic cell populations with short-term repopulating activity selectively engraft NOD/SCID-beta2 microglobulin-null mice. J Clin Invest 2001; 107:199-206
17. Goldschneider I, Komschlies KL, Greiner DL. Studies of thymocytopoiesis in rats and mice. I. Kinetics of appearance of thymocytes using a direct intrathymic adoptive transfer assay for thymocyte precursors. J Exp Med 1986; 163:1-17

18. Guenechea G, Gan OI, Dorrell C, Dick JE. Distinct classes of human stem cells that differ in proliferative and self-renewal potential. Nat Immunol 2001; 2:75-82
19. Hiramatsu H, Nishikomori R, Heike T, Ito M, Kobayashi K, Katamura K, Nakahata T, Complete reconstitution of human lymphocytes from cord blood CD34+ cells using the NOD/SCID/gammacnull mice model. Blood 2003; 102:873-880
20. Hogan CJ, Shpall EJ, Keller G. Differential long-term and multilineage engraftment potential from subfractions of human CD34+ cord blood cells transplanted into NOD/SCID mice. Proc Natl Acad Sci USA 2002; 99:413-418
21. Hou YH, Srour EF, Ramsey H, Dahl R, Broxmeyer HE, Hromas R. Identification of a human B-cell/myeloid common progenitor by the absence of CXCR4. Blood 2005; 105:3488-3492
22. Ito M, Hiramatsu H, Kobayashi K, Suzue K, Kawahata M, Hioki K, Ueyama Y, Koyanagi Y, Sugamura K, Tsuji K, Heike T, Nakahata T. NOD/SCID/γ_c^{null} mouse: an excellent recipient mouse model for engraftment of human cells. Blood 2002; 100:3175-3182
23. Javazon EH, Beggs KJ, Flake AW. Mesenchymal stem cells: paradoxes of passaging. Exp Hematol 2004; 32:414-425
24. Kawada H, Fujita J, Kinjo K et al. Nonhematopoietic mesenchymal stem cells can be mobilized and differentiate into cardiomyocytes after myocardial infarction. Blood 2004; 104:3581-3587
25. Kiel M, Ylimaz OH, Iwashita T, Ylimaz OH, Terhorst D, and Morrison SJ. SLAM family receptors distinguish hematopoietic stem and progenitor cells and reveal endothelial niches for stem cells. Cell 2005; 121:1109-1121
26. Koc ON, Gerson SL, Cooper BW et al. Rapid hematopoietic recovery after coinfusion of autologous-blood stem cells and culture-expanded marrow mesenchymal stem cells in advanced breast cancer patients receiving high-dose chemotherapy. J Clin Oncol 2000; 18:307-316
27. Larochelle A, Vormoor J, Hanenberg H, Wang JC, Bhatia M, Lapidot T, Moritz T, Murdoch B, Xiao XL, Kato I, Williams DA, Dick JE. Identification of primitive human hematopoietic cells capable of repopulating NOD/SCID mouse bone marrow: implications for gene therapy. Nat Med 1996; 2:1329-1337
28. Lemischka IR, Raulet DH, Mulligan RC. Developmental potential and dynamic behavior of hematopoietic stem cells. Cell 1986; 45:917-972
29. Majumdar MK, Thiede MA, Mosca JD, Moorman M, Gerson SL. Phenotypic and functional comparison of cultures of marrow-derived mesenchymal stem cells (MSCs) and stromal cells. J Cell Physiol 1998; 176:57-66
30. Matsuzaki Y, Kinjo K, Mulligan RC, Okano H. Unexpectedly efficient homing capacity of purified murine hematopoietic stem cells. Immunity 2004; 20:87-93
31. Mazurier F, Doedens M, Gan OI, Dick JE. Rapid myeloerythroid repopulation after intrafemoral transplantation of NOD-SCID mice reveals a new class of human stem cells. Nat Med 2003; 9:959-963
32. Mckenzie JI, Gan OI, Doedens M, Wang JCY, Dick JE. Individual stem cells with highly variable proliferation and self-renewal properties comprise the human hematopoietic stem cell compartment. Nat Immunol 2006; 11:1225-1233
33. Miyoshi H, Smith KA, Moiser DE, Verma IM, and Torbett BE. Transduction of human CD34+ cells that mediate long-term engraftment of NOD/SCID mice by HIV vectors. Science 1999; 283:682
34. Montecino-Rodriguez E, Leathers H, Dorshkind K. Bipotential B-macrophage progenitors are present in adult bone marrow. Nat Immunol 2001; 2:83-88
35. Muguruma Y, Yahata T, Miyatake H, Sato T, Uno T, Itho J, Itho M, Kato S, Hotta T, Ando K. Reconstitution of the functional human hematopoietic microenvironment derived from human mesenchymal stem cells in the murine bone marrow compartment. Blood 2006; 107:1878-1887
36. Nilsson SK, Johnston HM, Whitty GA et al. Osteopontin, a key component of the hematopoietic stem cell niche and regulator of primitive hematopoietic progenitor cells. Blood 2005; 106:1232-1239

37. Nolta JA, Dao MA, Wells S, Smogorzewska EM, Kohn DB. Transduction of pluripotent human hematopoietic stem cells demonstrated by clonal analysis after engraftment in immune-deficient mice. Proc Natl Acad Sci USA 1996; 93:2414-2419
38. Noort WA, Kruisselbrink AB, in 't Anker PS et al. Mesenchymal stem cells promote engraftment of human umbilical cord blood-derived CD34⁺ cells in NOD/SCID mice. Exp Hematol 2002; 30:870-878
39. Osawa M, Hanada K, Hamada H, Nakauchi H. Long-term lymphohematopoietic reconstitution by a single CD34-low/negative hematopoietic stem cell. Science 1996; 273: 242-245
40. Peled A et al. Dependence of human stem cell engraftment and repopulation of NOD/SCID mice on CXCR4. Science 1999; 283:845-848
41. Pittenger MF, Mackay AM, Beck SC et al. Multilineage potential of adult human mesenchymal stem cells. Science 1999; 284:143-147
42. Ponomaryov T, Peled A, Petit I et al. Induction of the chemokine stromal-derived factor-1 following DNA damage improves human stem cell function. J Clin Invest 2000; 106:1331-1339
43. Prockop DJ. Marrow stromal cells as stem cells for nonhematopoietic tissues. Science 1997; 276:71-74
44. Schmidt M, Hoffmann G, Wissler M, Lemke N, Mussig A, Glimm H, Williams DA, Ragg S, Hesemann CU, von Kalle C. Detection and direct genomic sequencing of multiple rare unknown flanking DNA in highly complex samples. Hum Gene Ther 2001; 12:743-749
45. Shultz LD et al. Multiple defects in innate and adaptive immunologic function in NOD/LtSz-scid mice. J Immunol 1995; 154:180-191
46. Shultz LD, Lyons BL, Burzenski LM, Gott B, Chen X, Chaleff S, Kotb M, Gillies SD, King M, Mangada J, Greiner DL, Handgretinger R. Human lymphoid and myeloid cell development in NOD/LtSz-scid IL2R gamma null mice engrafted with mobilized human hemopoietic stem cells. J Immunol 2005; 174:6477-6489
47. Stier S, Ko Y, Forkert R et al. Osteopontin is a hematopoietic stem cell niche component that negatively regulates stem cell pool size. J Exp Med 2005; 201:1781-1791
48. Sugiyama T, Kohara H, Noda M, Nagasawa T. Maintenance of the hematopoietic stem cell pool by CXCL12-CXCR4 chemokine signaling in bone marrow stromal cell niches. Immunity 2005; 25:977-988
49. van der Loo JC et al. Nonobese diabetic/severe combined immunodeficiency (NOD/SCID) mouse as a model system to study the engraftment and mobilization of human peripheral blood stem cells. Blood 1998; 92:2556-2570
50. Wagers AJ, Sherwood RI, Christensen JL, Weissman IL. Little evidence for developmental plasticity of adult hematopoietic stem cells. Science 2002; 297:2256-2259
51. Westen H, Bainton DF. Association of alkaline-phosphatase-positive reticulum cells in bone marrow with granulocytic precursors. J Exp Med 1979; 150:919-937
52. Yahata T, Ando K, Nakamura Y, Ueyama Y, Shimamura K, Tamaoki N, Kato S, Hotta T. Functional human T lymphocyte development from cord blood CD34+ cells in nonobese diabetic/Shi-scid, IL-2 receptor gamma null mice. J Immunol 2002; 169:204-209
53. Yahata T, Ando K, Sato T et al. A highly sensitive strategy for SCID-repopulating cell assay by direct injection of primitive human hematopoietic cells into NOD/SCID mice bone marrow. Blood 2003; 101:2905-2913
54. Yahata T, Yumino S, Sheng Y, Miyatake H, Uno T, Muguruma Y, Ito M, Miyoshi H, Kato S, Hotta T, and Ando K. Clonal analysis of thymus-repopulating cells presents direct evidence for self-renewal division of human hematopoietic stem cells. Blood 2006; 108:2446-2454
55. Zhang J, Niu C, Ye L et al. Identification of the haematopoietic stem cell niche and control of the niche size. Nature 2003; 425:836-841

The Differentiative and Regenerative Properties of Human Hematopoietic Stem/Progenitor Cells in NOD-SCID/IL2rγ^{null} Mice

F. Ishikawa(✉), Y. Saito, S. Yoshida, M. Harada, and L. D. Shultz

1 Introduction 88
2 Newborn Immune-Compromised Mice as Recipients 88
   2.1 Newborn scid Mice as Recipients 88
   2.2 Human Immune Reconstitution in Engrafted NOD-SCID/IL2rγ^{null} Mice 89
3 Studying Stem Cell Plasticity in the Humanized Mouse 89
   3.1 Background 89
   3.2 Human Epithelial Cells in the Gastrointestinal Tissues of Humanized Mice 90
   3.3 Human Insulin$^+$ Cells in the Pancreas of Humanized Mice 90
   3.4 Human HSC-Derived Cardiomyocytes in the Cardiac Tissue of Humanized Mice 91
   3.5 Differentiative Capacity of Human Mesenchymal Stem Cells in scid Mice 92
4 Conclusion 93
References 93

Abstract Biomedical research including immunology and stem cell biology has developed greatly because of the evolving technology of gene modification and conventional transplantation methods using the most common experimental laboratory animal, the mouse. To translate promising research findings based on mouse research into clinical medicine, however, we need to clarify whether similar events take place in humans. In the study of hematology and immunology, humanized mice provide a unique and efficient experimental system to evaluate differentiation, function, and interaction of human blood cells or immune components. Here we review the latest experimental findings in the fields of immunology, stem cell biology, and regenerative medicine using humanized mice.

F. Ishikawa
Research Unit for Human Disease Model, RIKEN Research Center for Allergy and Immunology, 1-7-22 Suehiro-cho Tsurumi-ku, Yokohama, Kanagawa 230-0045, Japan
f_ishika@rcai.riken.jp

Abbreviations BM: bone marrow; CB: cord blood; FACS: fluorescence-activated cell sorter; GFP: green fluorescent protein; HSC: hematopoietic stem cell

1 Introduction

In vivo analysis of physiological and pathophysiological processes is often crucial to understand the mechanisms governing these processes. In the past, much has been gained by the study of normal physiology and of disease processes mimicking human pathology in mice, by taking advantage of naturally occurring murine diseases and genetically modified mouse models. However, the mechanisms operating in the murine system are not necessarily identical to those in the human system, and ethical and technical limitations make the in vivo study of human physiology and pathophysiology difficult. To overcome these constraints, a number of xenotransplantation systems using various immune-compromised mouse strains have been developed.

2 Newborn Immune-Compromised Mice as Recipients

2.1 *Newborn scid Mice as Recipients*

Throughout the history of human cell research using immune-compromised mice, the vast majority of investigators have utilized adult mice as recipients (McCune et al. 1988; Shultz et al. 2005; Kollet et al. 2000). We and others have reported that highly efficient engraftment of human cord blood (CB) cells is achieved when they are transplanted during the neonatal phase of NOD.Cg-$Prkdc^{scid}B2m^{tm1Unc}$ (NOD-scid/β2m^{null}) or BALB/c-$Rag2^{null}$ $IL2r\gamma^{null}$ (RAG2$^{-/-}$/$\gamma_c^{-/-}$) mice (Ishikawa et al. 2002; Traggiai et al. 2004). In addition to the increased cell dose per body weight in neonatal recipients, the immature microenvironment may be more permissive and supportive for xenogeneic human stem/progenitor cell engraftment and reconstitution compared with adult recipients. Optimization of the homing efficiency of human hematopoietic stem cells (HSCs) is essential for development of highly chimeric humanized mice. Traggiai et al. injected purified CD34$^+$ CB cells into the liver of newborn RAG2$^{-/-}$/$\gamma_c^{-/-}$ mice, taking advantage of the fact that the liver acts as a major hematopoietic organ during the fetal and neonatal periods. For utilization of the bone marrow (BM) niche, we injected purified human hCD34$^+$hCD38$^-$ CB cells into newborns via the facial vein. In NOD.cg-$Prkdc^{scid}Il2rg^{tm1Wjll}$ (NOD-SCID/IL2rγ^{null}) recipients, the levels of human hematopoietic cell chimerism were significantly higher than in NOD-scid/β2m^{null} mice. Importantly, in both the intrahepatic (i.h.) RAG2$^{-/-}$/$\gamma_c^{-/-}$ mouse transplantation and the intravenous (i.v.) NOD-SCID/

IL2rγ^{null} mouse transplantation models, significant levels of human chimerism persisted for >24 weeks after transplantation, and hCD34$^+$ BM cells derived from primary recipients reconstituted hematopoiesis in secondary recipients, demonstrating that these systems support the self-renewal of human HSCs. In the reconstituted mice, myelo-monocytes, dendritic cells, erythroid cells, platelets, and T-, B-, and NK-lymphocytes were detected, demonstrating the multipotential capacity of human HSCs.

2.2 Human Immune Reconstitution in Engrafted NOD-SCID/IL2rγ^{null} Mice

To evaluate the function of the reconstituted human adaptive immunity in mice, we analyzed the production of human immunoglobulins in sera of the engrafted mice by ELISA. In the sera from all NOD/SCID/IL2rγ^{null} recipients, significant amounts of IgG (~100 μg/ml) and IgM (~300 μg/ml) were detectable, whereas sera from the engrafted NOD/SCID/β2m$^{-/-}$ mice contained lower levels of IgM (<100 μg/ml) and little or no IgG. These data collectively suggest that class switching occurs effectively in NOD/SCID/IL2rγ^{null} mice. Additionally, when the engrafted mice were immunized with a T-dependent antigen, significant levels of Ag-specific human IgM and IgG were detected in all immunized mice sera but not from nonimmunized engrafted mice sera, demonstrating that the reconstituted human adaptive immune system functions properly in the NOD/SCID/IL2rγ^{null} recipients to produce antigen-specific human IgM and IgG antibodies. Comparable amounts of human immunoglobulins were produced in the newborn RAG2$^{-/-}$/γ$_c^{-/-}$ mouse transplantation model as well.

3 Studying Stem Cell Plasticity in the Humanized Mouse

3.1 Background

At the very end of the twentieth century, Petersen et al. reported that BM cells possess an extensive capacity to generate hepatocytes as well as hematopoietic cells (Petersen et al. 1999). The notion that BM-derived progenitor cells are capable of generating nonhematopoietic epithelial cells was confirmed by other reports (Harris et al. 2004; Wang et al. 2005). To translate the multipotential capacity of BM HSCs into a clinical setting, this capacity must be tested with human cells. By restoring the age-related decline of tissue progenitor activity through exposure to environmental factors secreted by juvenile hosts, the immature environment of the neonatal recipients may potentiate the developmental plasticity of the transplanted cells. The newborn scid-repopulation model would then be expected to be a sensitive assay for detecting very rare regenerating events and for identifying the mechanism underlying regeneration from human stem cells in vivo.

3.2 Human Epithelial Cells in the Gastrointestinal Tissues of Humanized Mice

Krause et al. demonstrated that a single hematopoietic cell is able to give rise to intestinal epithelial cells and alveolar epithelial cells in a mouse syngeneic transplantation system (Krause et al. 2001). Today, regenerative medicine has become accepted as a potential novel therapy for the twenty-first century (Kassem 2006; Lagasse et al. 2001; Lindvall and Kokaia 2006; Prockop et al. 2003). Whether human HSCs possess a similar capacity to give rise to nonhematopoietic cells and tissues is controversial. We aimed to elucidate whether human hematopoietic stem/progenitor cells possess the capacity to generate epithelial cells in vivo. When we analyzed NOD/SCID/$\beta 2m^{-/-}$ mice transplanted with human CB-derived CD34$^+$ cells, human chromosome$^+$ epithelial cells were detected in the murine gastrointestinal tissues. In addition to quantitative analysis of the incidence of the epithelial cells of human origin, xenogeneic transplantation enabled us to examine the possibility of cell fusion as a mechanism for the generation of the donor-derived epithelial cells by double FISH analyses using probes specific for murine and human chromosomes. The findings from these analyses suggest that human hematopoietic tissues contained the stem/progenitor cells providing the capacity to generate gastrointestinal epithelial cells, albeit with a low efficiency. The frequency of epithelial cell regeneration was as low as 0.1%-0.3%, which may be suboptimal to regenerate an extensive area of epithelial loss. However, immunological disturbance is thought to be involved in the pathophysiology of inflammatory bowel disorders such as Crohn's disease and ulcerative colitis. Transplantation of hematopoietic tissue-derived progenitors may result in the normalization of the host immune system as well as the repair of injured epithelial cells. HSC transplantation, by exploiting the multipotential capacity of human CD34$^+$ cells in generating epithelial cells as well as immuno-hematopoietic cells, may become one of the therapies for intractable bowel disorders in the future.

3.3 Human Insulin$^+$ Cells in the Pancreas of Humanized Mice

The regeneration of pancreatic beta cells has been long awaited as the curative therapy for insulin-dependent and non-insulin-dependent diabetes mellitus. Although pancreatic organ transplantation and pancreatic islet transplantation have been performed in end-stage diabetic patients, the scarcity of donor tissues is an obstacle that has yet to be resolved. In the past, several reports described the regeneration of BM-derived pancreatic beta cells based on mouse syngeneic or allogeneic transplantation assays (Camargo et al. 2004). Transplantation of mouse BM mononuclear cells has been reported to result in the generation of donor-derived insulin+ cells in the recipient pancreas islet (Hess et al. 2003). On the other hand, Ianus et al. suggested that the generation of endothelial cells in the pancreas indirectly enhanced the generation of host-derived beta cells, leading to the functional amelioration of streptozotocin-induced glucose intolerance in mice (Ianus et al. 2003). We investigated the regenerative property of human hematopoietic tissue-derived cells and obtained insights into mechanisms underlying regeneration of insulin-producing cells with the newborn NOD-SCID/$\beta 2m^{null}$ xenotransplantation model. The incidence of human CB-derived insulin+ cells was $0.65 \pm 0.64\%$ ($n=6$) in this system. We confirmed human insulin production in the recipient pancreatic tissue but not in the pancreas of nontransplanted control NOD-SCID/$\beta 2m^{null}$ mice by RT-PCR. Dual FISH analyses using species-specific probes demonstrate the presence of human chromosome+ mouse chromosome+ insulin+ cells and human chromosome+ mouse chromosome- insulin+ cells at a similar frequency. The presence of nearly equal proportions of human chromosome+ murine chromosome- cells and human chromosome+ murine chromosome+ cells indicates that both differentiation and cell fusion contribute to the generation of donor marker+ insulin-producing cells following human CB cell transplantation into xenogeneic hosts. Along with the reports from studies using mouse syngeneic transplantations, the in vivo production of human insulin-producing cells may encourage the future utilization of regenerative medicine in the treatment of diabetes mellitus. In the future, the human cells responsible for the regeneration of insulin-producing cells via differentiation or via cell fusion must be identified. Furthermore, it will be necessary to address the question of whether the efficiency of insulin+ cell regeneration in humanized mice is increased after pancreatic injury induced by streptozotocin.

3.4 *Human HSC-Derived Cardiomyocytes in the Cardiac Tissue of Humanized Mice*

While c-Kit+ cells and Sca-1+ cells reside in cardiac tissue as candidate tissue progenitors, whether hematopoietic tissue-derived cells regenerate cardiomyocytes has been controversial (Balsam et al. 2004; Murry et al. 2004; Nygren et al. 2004; Orlic et al. 2001a, 2001b). We therefore set out to investigate the in vivo generation of cardiomyocytes from murine and human hematopoietic tissues with the neonatal NOD-SCID/IL2rγ^{null} transplantation model. Transplantation of Lin-Sca1+ cells resulted in the generation of cardiomyocytes in the recipient mice. When we used GFP+ BM cells as donors and CFP transgenic newborn mice as recipients, GFP+ cardiomyocytes coexpressed CFP in the cytoplasm, indicating that cellular fusion between donor and host cells had occurred. To further examine the in vivo regenerative capacity of human hematopoietic stem cells, $2\text{-}5 \times 10^4$ Lin-hCD34+hCD38- FACS-purified human HSCs from human CB mononuclear cells were intravenously injected into neonatal NOD/SCID/IL2rγ^{null} mice. At 2-4 months after

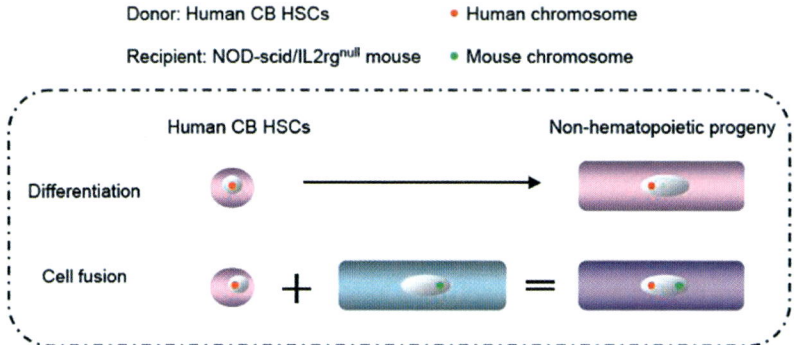

Fig. 1 Dual FISH analysis allows discrimination between direct regeneration and cell fusion as the mechanism underlying human stem cell-derived progeny in mouse tissues. Probes for human chromosomes and for mouse chromosomes were used to identify whether a cell is of donor or recipient origin, or both

transplantation, the cardiac tissues of these immunologically humanized mice were analyzed for the presence of human chromosome-containing cardiomyocytes. The human chromosome-containing cardiac cells expressed connexin 43 on their periphery and exhibited cardiomyocyte-specific striations. Serial confocal imaging was used to pinpoint signals for human and murine chromosomes within single cells and enabled us to exclude the possibility that they arose from cell overlay. To clarify the origin of each cardiomyocyte, we performed dual FISH analysis on the cardiac tissues derived from the recipient mice (Fig. 1). With this method, the presence of murine chromosomes was demonstrated in all human chromosome-containing cardiomyocytes. This finding suggests that the transplantation of human HSCs resulted in the generation of donor maker+ cardiomyocytes through cellular fusion between donor-derived hematopoietic progeny and host cardiomyocytes, not through transdifferentiation.

3.5 Differentiative Capacity of Human Mesenchymal Stem Cells in scid Mice

In addition to the analysis of immunologically humanized mice developed by the transplantation of human HSCs, several investigators have reported the differentiative capacity of human mesenchymal stem cells (MSCs) in vivo. Sato et al. demonstrated the generation of hepatocytes by human MSC transplantation in the NOD-scid model (Sato et al. 2005). Their findings suggested that MSCs did not fuse with hepatocytes but directly differentiated into hepatocytes. Toma et al. reported the capacity of human MSCs to give rise to cardiomyocytes in CB17-scid mice (Toma et al. 2002). In this system, again, human MSCs were shown to generate cardiomyocytes via a fusion-independent pathway.

4 Conclusion

The multipotential properties and self-renewal capacity of hematopoietic tissue-derived stem cells have been successfully analyzed in the humanized mouse model. These findings, to some extent, confirmed the findings that have been demonstrated with mouse BM stem cells. The plasticity of human HSCs may be explained in part by the fusion between the stem cells and nonhematopoietic cells, which is a mechanism that has been demonstrated in the murine system with sex-mismatch syngeneic transplantation experiments. To translate stem cell plasticity into regenerative medicine, the contribution of donor cells to regeneration of the target tissue must be clarified with each disease model. Current studies show that the frequency of regeneration of nonhematopoietic cells through human hematopoietic stem/progenitor cell transplantation is quite low. It is possible to hypothesize that higher regeneration efficiencies leading to improvement in functional improvement may be observed in the settings of specific injury or disease. By creating in vivo models for human diseases with neonatal xenotransplantation, it will be possible to test such hypotheses. While NOD/SCID/IL2rγ^{null} mice are highly efficient xenograft recipients because of the high levels of suppression of both natural and adaptive immunity, there exists a clear incompatibility in major histocompatibility complex between the donor and the recipient. That the model has proven to be useful for studying human stem cell plasticity and tissue regeneration even with such an immunological barrier indicates the potential of humanized mice as an effective tool for future investigations in regenerative therapies for human diseases.

References

Balsam, L. B., Wagers, A. J., Christensen, J. L., Kofidis, T., Weissman, I. L., and Robbins, R. C. (2004). *Nature* **428,** 668-73.
Camargo, F. D., Finegold, M., and Goodell, M. A. (2004). *J Clin Invest* **113,** 1266-70.
Harris, R. G., Herzog, E. L., Bruscia, E. M., Grove, J. E., Van Arnam, J. S., and Krause, D. S. (2004). *Science* **305,** 90-3.
Hess, D., Li, L., Martin, M., Sakano, S., Hill, D., Strutt, B., Thyssen, S., Gray, D. A., and Bhatia, M. (2003). *Nat Biotechnol* **21,** 763-70.
Ianus, A., Holz, G. G., Theise, N. D., and Hussain, M. A. (2003). *J Clin Invest* **111,** 843-50.
Ishikawa, F., Livingston, A. G., Wingard, J. R., Nishikawa, S., and Ogawa, M. (2002). *Exp Hematol* **30,** 488-94.
Kassem, M. (2006). *Ann NY Acad Sci* **1067,** 436-42.
Kollet, O., Peled, A., Byk, T., Ben-Hur, H., Greiner, D., Shultz, L., and Lapidot, T. (2000). *Blood* **95,** 3102-5.

Krause, D. S., Theise, N. D., Collector, M. I., Henegariu, O., Hwang, S., Gardner, R., Neutzel, S., and Sharkis, S. J. (2001). *Cell* **105,** 369-77.

Lagasse, E., Shizuru, J. A., Uchida, N., Tsukamoto, A., and Weissman, I. L. (2001). *Immunity* **14,** 425-36.

Lindvall, O., and Kokaia, Z. (2006). *Nature* **441,** 1094-6.

McCune, J. M., Namikawa, R., Kaneshima, H., Shultz, L. D., Lieberman, M., and Weissman, I. L. (1988). *Science* **241,** 1632-9.

Murry, C. E., Soonpaa, M. H., Reinecke, H., Nakajima, H., Nakajima, H. O., Rubart, M., Pasumarthi, K. B., Virag, J. I., Bartelmez, S. H., Poppa, V., Bradford, G., Dowell, J. D., Williams, D. A., and Field, L. J. (2004). *Nature* **428,** 664-8.

Nygren, J. M., Jovinge, S., Breitbach, M., Sawen, P., Roll, W., Hescheler, J., Taneera, J., Fleischmann, B. K., and Jacobsen, S. E. (2004). *Nat Med* **10,** 494-501.

Orlic, D., Kajstura, J., Chimenti, S., Bodine, D. M., Leri, A., and Anversa, P. (2001a). *Ann NY Acad Sci* **938,** 221-9; discussion 229-30.

Orlic, D., Kajstura, J., Chimenti, S., Jakoniuk, I., Anderson, S. M., Li, B., Pickel, J., McKay, R., Nadal-Ginard, B., Bodine, D. M., Leri, A., and Anversa, P. (2001b). *Nature* **410,** 701-5.

Petersen, B. E., Bowen, W. C., Patrene, K. D., Mars, W. M., Sullivan, A. K., Murase, N., Boggs, S. S., Greenberger, J. S., and Goff, J. P. (1999). *Science* **284,** 1168-70.

Prockop, D. J., Gregory, C. A., and Spees, J. L. (2003). *Proc Natl Acad Sci USA* **100 Suppl 1,** 11917-23.

Sato, Y., Araki, H., Kato, J., Nakamura, K., Kawano, Y., Kobune, M., Sato, T., Miyanishi, K., Takayama, T., Takahashi, M., Takimoto, R., Iyama, S., Matsunaga, T., Ohtani, S., Matsuura, A., Hamada, H., and Niitsu, Y. (2005). *Blood* **106,** 756-63.

Shultz, L. D., Lyons, B. L., Burzenski, L. M., Gott, B., Chen, X., Chaleff, S., Kotb, M., Gillies, S. D., King, M., Mangada, J., Greiner, D. L., and Handgretinger, R. (2005). *J Immunol* **174,** 6477-89.

Toma, C., Pittenger, M. F., Cahill, K. S., Byrne, B. J., and Kessler, P. D. (2002). *Circulation* **105,** 93-8.

Traggiai, E., Chicha, L., Mazzucchelli, L., Bronz, L., Piffaretti, J. C., Lanzavecchia, A., and Manz, M. G. (2004). *Science* **304,** 104-7.

Wang, G., Bunnell, B. A., Painter, R. G., Quiniones, B. C., Tom, S., Lanson, N. A., Jr., Spees, J. L., Bertucci, D., Peister, A., Weiss, D. J., Valentine, V. G., Prockop, D. J., and Kolls, J. K. (2005). *Proc Natl Acad Sci USA* **102,** 186-91.

Antigen-Specific Antibody Production of Human B Cells in NOG Mice Reconstituted with the Human Immune System

R. Ito, M. Shiina, Y. Saito, Y. Tokuda, Y. Kametani, and S. Habu(✉)

1	Introduction	96
2	Antibody Production in the CB-NOG Mouse After Immunization	97
	2.1 Past Studies on Human Antibody Production in Immunodeficient Mice	97
	2.2 Antigen-Specific IgM Antibody is Dominant in Immunized CB-NOG Mice	98
	2.3 Human CD34+Cells Preferentially Develop into CD5+B Cells in the CB-NOG Spleen	99
3	Limited Function of Human T Cells Developed in the Murine Thymus	99
	3.1 IL-2 Production is Defective in T Cells Activated with Immunized Antigen in CB-NOG Mice	99
	3.2 Human T Cells Are Positively Selected by Murine MHC in the NOG Thymus	101
4	Antigen-Specific IgG Antibody in NOG Mice with Human Lymph Node Engraftment	102
5	Conclusion	105
References		106

Abstract Passive antibody administration shows strong potential as a new therapeutic method. In clinical applications, human-derived antibodies with antigen specificity are more useful without putting individuals at risk. Production of human-derived antibodies against given antigens can be obtained from animal models if the human immune system is established in the animals. In fact, past reports revealed that human T and B cells develop from hematopoietic progenitor cells in immunodeficient mice. However, there have been few reports on sufficient induction of antigen-specific antibodies, particularly IgG, in immunodeficient mice reconstituted with human immune cells. In this chapter, we discuss a major

---

S. Habu
Department of Immunology, Tokai University School of Medicine, Bohseidai, Isehara,
Kanagawa, 259-1193, Japan
sonoko@is.icc.u-tokai.ac.jp

shortcoming of induction of antigen-specific IgG antibodies in human immune cells developed in the murine environment based on our data. We demonstrated that human T cell development is restricted by the murine MHC and consequently T cells may not achieve cognate interaction with human B cells. Human B cells developed in the mouse are mainly CD5+B1 cells that preferentially produce IgM. At the same time, human LN transplantation on the spleen enabled NOG mice to produce antigen-specific IgG antibody. These results suggest that if efficient cognate interaction mediated by a certain antigen on MHC class II between human T and B-2 cells occurs, human B cells can produce IgG antibody against a given antigen in the murine environment.

Abbreviations APC: antigen-presenting cell; BM: bone marrow; CB: cord blood; DC: dendritic cell; DN: double negative; DNP-KLH: 2,4-dinitrophenylated keyhole limpet hemocyanin; DP: double positive; ELISA: enzyme-linked immunosorbent assay; FACS: fluorescence-activated cell sorting; GVHD: graft-versus-host disease; Ig: immunoglobulin; IL: interleukin; KO: knockout; LN: lymph node; mAb: monoclonal antibody; MACS: magnetic cell sorting; MHC: major histocompatibility complex; NK: natural killer; NOD: nonobese diabetic; NOG: NOD/Shi scid IL2R gamma chain knockout; OVA: ovalbumin; PBL: peripheral blood lymphocyte; PMA: phorbol 12-myristate 13-acetate; RTOC: reaggregate thymic organ culture; SCID: severe combined immunodeficiency; TSST-1: toxic shock syndrome toxin-1; TCR: T cell receptor; WT: wild type

1 Introduction

Reports showing the therapeutic effect of passive antibody administration in some diseases such as cancer and autoimmune diseases are increasing [4, 19, 24]. To date, most antibodies used for clinical application have been derived from mouse monoclonal antibodies (mAb) and humanization of the mAb has been performed with genetic engineering techniques. However, it is still uncertain whether patients who receive such humanized mAb suffer from certain harmful effects because they may produce their own antibody against administered mAb even if mAb are partially or almost completely humanized. Currently, some mutant mice have been highlighted as a tool for generating less risky mAb. Since these mutant lines carry human chromosome fragments containing immunoglobulin gene clusters [26], the gene product is identical to the human product. However, it is still possible that patients may produce antibody against mAb obtained from such mutant mice because their immunoglobulin contains murine-origin sugar components [15]. If we can develop a laboratory animal in which human-derived B cells can produce antibodies against given antigens, the antibodies should be extremely safe for clinical application. In the 1990s, investigators attempted to reconstitute human-origin hematopoiesis and/or lymphopoiesis by using severe combined immunodeficiency (SCID) mice that can receive xenografts because of their lack of T and B cell development [12, 16]. In those studies, however, human T cells did not develop in the mice if the human fetal thymus was not simultaneously engrafted [14,17] despite the fact that human T cells develop from cord blood (CB) or bone marrow (BM) cells in organ cultures of murine thymus lobes [20, 21, 28]. After other SCID background lines with no natural killer (NK) cells (NOD-SCID-IL-2Rg−/−, abbreviated as NOG mice) or with reduced NK activity (NOD-SCID mice) became available, in vivo development of human T [11, 27] and B [8, 13] cells from human CB and BM cells was reported in the mouse. Because of their remarkably high efficiency for normal human cell engraftment including CB cells [7], we have used NOG mice reconstituted with human CD34+CB cells (abbreviated as CB-NOG mice) to induce antigen-specific antibody derived from human B cells after immunization. Here, based on new and previous data concerning development of human T and B cells in NOG mice, we discuss why antigen-specific IgG antibody is almost undetectable in mice reconstituted with human CD34+ cells.

2 Antibody Production in the CB-NOG Mouse After Immunization

2.1 Past Studies on Human Antibody Production in Immunodeficient Mice

Initial studies attempted to induce antibody production from human B cells in immunodeficient mice such as SCID or NOD-SCID mice implanted with human peripheral blood lymphocytes (PBL) containing mature T and B cells [1, 16, 22]. However, these attempts were almost all unsuccessful, presumably because graft-versus-host diseases (GVHD) occurred because of the presence of human immune cells in the recipient mice. To avoid GVHD the engrafted PBL number was decreased and the problem of GVHD was basically eliminated, but the antigen-specific antibody became almost undetectable, presumably because human immune cells are dispersed at a low density in mouse lymphoid tissues, resulting in insufficient cell-to-cell interaction of the immune cells as discussed by Sandhu et al. [22].

To overcome these problems, we implanted human CD34+CB cells into NOG or NOD-SCID mice to induce the development of human immune cells that adapt to the murine environment. In these reconstituted mice, considerable numbers of T and B cells develop and accumulate in the murine lymphoid tissues [5, 25]. To avoid GVHD, we are careful to purify CD34+CB cells before their implantation by double-positive selection using MACS beads and FACS [8], and have succeeded in efficient reconstitution of human T and B cells by CD34+ cells derived from CB, BM, and mobilized PBL.

2.2 Antigen-Specific IgM Antibody is Dominant in Immunized CB-NOG Mice

To determine whether antigen-specific antibody is produced in CB-NOG mice after immunization, CD34+CB cells purified by the double selection process were implanted into irradiated NOG mice. After development of human T and B cells was confirmed by marking the expression of CD3 and CD19/IgM, respectively, in peripheral blood, CB-NOG mice were immunized six times with DNP-KLH every 2 weeks. Figure 1 shows representative results of the raised antibody in immunized CB-NOG mice. Total immunoglobulin levels of IgM and IgG in the serum were equally increased after immunization. However, the serum level of antigen-specific IgG (anti-DNP-KLH) was detected but was extremely low, if detectable, in comparison with the IgM. Moreover, only two of nine mice had detectable IgG, while specific IgM was found in all experimental mice. In addition to hapten-carrier antigen, similar results were obtained when CB-NOG mice were immunized with other antigens such as OVA, OVA peptide, and superantigen (data not shown). Ishikawa et al. [6] reported similar results showing that specific IgG against OVA is detectable in NOG mice if the newborn mice are transplanted with human stem cells, although the antibody levels are very low. These results indicate that the CB-NOG mouse environment permits human B cell development to spontaneously produce IgG as well as IgM but may not be able to induce antigen-specific IgG antibody, which will be discussed later.

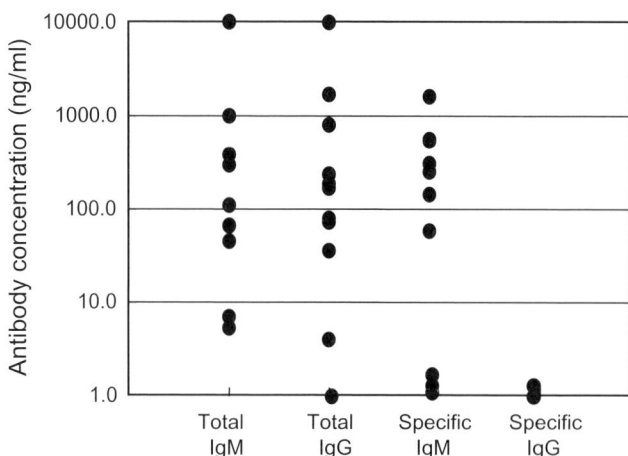

Fig. 1 Antibody level of total and antigen-specific IgM and IgG in immunized CB-NOG mice. NOG mice ($n=11$) were immunized with DNP-KLH emulsified with alum 6 times biweekly after implantation of human CD34+CB cells. After the last immunization at the 10th week, the concentration of anti-DNP-KLH in the mouse serum was measured by ELISA

2.3 Human CD34+Cells Preferentially Develop into CD5+B Cells in the CB-NOG Spleen

We have demonstrated that more than 50% of CD19+ cells in the CB-NOG mouse are CD5 positive but CD3 negative. This CD5 predominance did not change when CB, BM, or PBL was used as the CD34+ cell source [13]. These CD5+ B cells also expressed IgM, IgD, and CD20 in addition to CD19, indicating that human CD34+ cells develop into mature type CD5+B cells in the NOG environment. CD5+B cells belong to a different B cell subpopulation, termed B1 cells, that are distinguished from ordinary B cells, termed B2 cells [9, 10, 18]. At the same time, CD5+B cells are known to produce mainly IgM but are less likely to produce antigen-specific IgG. Such characteristics of CD5+B cells may provide a clue to explain why antigen-specific IgG human antibody is not raised very much in immunized CB-NOG mice.

Why do human CD5+B cells develop with high efficiency in NOG mice? Under physiological conditions, the proportion of CD5 1 B cells is known to be low in PBL and spleen, and CD5+B cells are predominantly located in the peripheral cavities of mice [25]. Moreover, CD5+B cells are rare in human tissues. Matsumura et al. [13] found that the proportion of CD5+B cells increased in the NOG spleen with time after transplantation but was low without an increase in BM. In fact, CD5-IgM-CD19+ cells became CD5+B cells in cocultures with nonreconstituted NOG spleen cells. In light of these findings, it is possible to speculate that the murine spleen environment may possess the potential to force human CD34+ cells to become the CD5+B cell lineage or may induce CD5+B cell proliferation and/or accumulation. At present, we do not have any suitable evidence to support either possibility, but our findings should provide tools for studying the development of human CD5+B cells, which is still controversial.

3 Limited Function of Human T Cells Developed in the Murine Thymus

3.1 IL-2 Production is Defective in T Cells Activated with Immunized Antigen in CB-NOG Mice

Since T cell help is required for a class switch of immunoglobulin from IgM to IgG after immunization, one may ask the question of whether human T cells developed in the CB-NOG mouse possess helper function in NOG mice in the same way as they develop physiologically in the human thymic environment. Past reports including ours demonstrated that human CD34+CB cells develop into CD4 and CD8 single-positive (CD4+ and CD8+) cells through CD4−8− double-negative (DN) cells and CD4+8+ double-positive (DP) cells in the murine thymic environment in vitro [21, 28] and in vivo [8, 21]. Single-positive thymocytes expressed mature type

T cell markers such as CD1a low and TCR high, and possessed the potential to produce cytokines after stimulation with the mitogen PMA/ionomycin [21]. In CB-NOG peripheral lymphoid tissues, human T cells developed from CD34+CB cells also express mature T cell markers [8].

In this study, we examined the cytokine-producing ability of human T cells after they have left the thymus in CB-NOG mice. When CD3high+ cells isolated from the CB-NOG spleen were stimulated with anti-CD3 or PMA/ionomycin in vitro, they expressed high levels of CD25, CD69, and CD154 (CD40L) and produced IL-2 in the culture supernatants (Fig. 2a, b), indicating that human T cells as well as mature thymocytes are activated to produce cytokine by TCR-mediated signaling. However, when CD3high+ T cells were isolated from the spleen of a CB-NOG mouse immunized with DNP-KLH and were cultured with nonreconstituted NOG spleen cells in the presence of the same antigen, their IL-2 production ability (Fig. 2c) and cell proliferation ability (data not shown) were extremely low in comparison with those

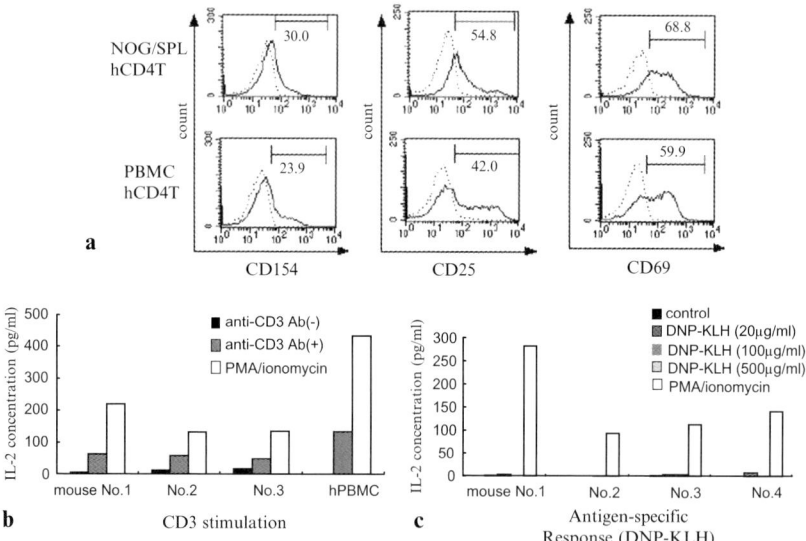

Fig. 2 Human T cells developed in CB-NOG mice are not activated by conventional antigens but activated by TCR-crosslinking. **a, b** T cells collected from CB-NOG spleen without stimulation (NOG/SPL hCD4T) or from PBL (PBMC hCD4T) of healthy volunteers were stimulated with anti-human CD3 antibody or PMA/Ionomycin. The in vitro stimulated T cells were subjected to flow cytometric analysis to determine the expression of activated antigens CD154 (16-h culture), CD25 and CD69 (48-h culture) (**a**). The concentration of human IL-2 in the culture supernatants was measured by ELISA. *Closed squares*, cells without anti-CD3 treatment. *Gray squares*, cells with anti-CD3 treatment. *Open squares*, cells treated with PMA/ionomycin (**b**). **c** CB-NOG mice were immunized with DNP-KLH emulsified with alum biweekly. One week after the 3rd booster, CD4 T cells collected by a magnetic bead system were cocultured with mitomycin C-treated non-immunized spleen cells without T cells in the presence of serially diluted DNP-KLH (20-500 μg/ml) for 48-h. The IL-2 concentration of the cultured supernatants was measured by ELISA. *Control cells*, cells cultured without stimulation

of the stimulated T cells with PMA/ionomycin or superantigen TSST-1. Why do T cells in the immunized mouse not respond properly to a given antigen such as DNP-KLH, despite the fact that human T cells appeared in the CB-NOG spleen and they can be induced to produce cytokine through the TCR signaling pathway in vitro?

3.2 Human T Cells Are Positively Selected by Murine MHC in the NOG Thymus

One possible interpretation of the reduced response to T cells by the immunized mice is that human T cells developed in the NOG mouse do not appropriately recognize antigen peptide presented by the human MHC on human antigen-presenting cells (APC) because human T cells are positively selected by the murine MHC in the NOG thymus. To confirm this assumption, we examined whether or not human T cells substantially develop in the NOG thymus under the restriction of the murine MHC. For this purpose, we used a reaggregate thymic organ culture (RTOC) system [2, 23] in which hematopoietic stem cells are cocultured with thymic stromal cells by generating a three-dimensional structure. In RTOC, human CD34+CB cells were cocultured with thymic stromal cells obtained from the fetal thymus of MHC class II (I-Ab) knockout (KO) mice or wild-type (WT) C57BL/6 mice to generate human/murine hybrid culture clusters. The clusters were then engrafted beneath the NOG kidney capsule, and 8 weeks later cells in the clusters were subjected to flow cytometric analysis. The developing cells in RTOC showed that the proportions and total cell numbers of human CD4+cells were reduced in KO mouse-derived hybrid clusters in comparison with WT-derived clusters while CD8+ cells were intact (Fig. 3).

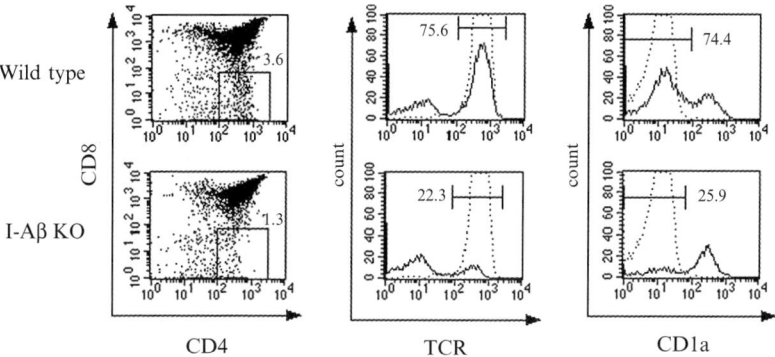

Fig. 3 Human T cells are positively selected by the murine MHC. Fetal thymic stromal cells obtained from C57BL/6 or I-Ab KO mice were cocultured with human CD34+CB cells to generate human/murine hybrid clusters. One week after reaggregation culture, the hybrid cluster was transplanted beneath the kidney capsule of NOG mice. Eight or 10 weeks after transplantation, lymphoid cells collected from the reaggregated clusters were stained with antibodies against human CD45, CD4, CD8, CD1a and TCR for flow cytometric analysis

Moreover, these CD4+ cells obtained from KO mouse-derived hybrid clusters expressed high CD1a+ and low CD3 (immature T cell markers) at relatively high rates. These results indicate that the majority of human T cells are positively selected to reach maturity under the restriction of the murine MHC in the NOG thymus.

The CB-NOG spleen may contain large numbers of human-derived hematopoietic cells but less murine cells because NOG mice received irradiation before CB cell implantation. Moreover, no B cell is present in NOG mice. Thus, human T cells developed under murine MHC restrictions have little chance to encounter murine APC including B cells for accomplishing cognate interaction for activation as memory cells. At the same time, these human T cells can not interact with human B cells and DC because they express human MHC. Collectively, it is strongly suggested that human T cells in the immunized CB-NOG mice respond poorly to in vitro restimulation with the immunized antigen because most of their TCR repertoire is restricted to murine MHC.

4 Antigen-Specific IgG Antibody in NOG Mice with Human Lymph Node Engraftment

When human T cells develop in the context of murine MHC restriction in CB-NOG mouse as mentioned above, they are unlikely to show suitable interaction with B cells expressing human MHC in CB-NOG mice, resulting in the failure of memory T cell development, which is responsible for the class switch from IgM to IgG. The ideal method to overcome the impediment of MHC restriction is to establish a system in which human T cells develop selectively in the murine thymic environment expressing human MHC. If the MHC gene of the NOG mice is replaced with the human MHC gene, T cell repertoires will be positively selected under human MHC restriction even in the NOG thymus. In such a mutant mouse, developed human T cells can be stimulated with given antigens that are presented by human MHC class II, resulting in induction of IgG antibody production in B cells through matched TCR-MHC interaction. This trial is now in progress.

Another system in which the human immune system can function under the same MHC restriction in laboratory animals is the engraftment of human lymphoid tissues in the immunodeficient mouse. In the initial studies, human embryonic thymus was implanted with human hematopoietic stem cells or embryonic hematolymphoid tissues in SCID mice, typical immunodeficient animals [14, 17]. This experimental system may enable T cells to interact with APC and with B cells via the human MHC in NOG mice, but it is not practical because human fetal tissues cannot be obtained because of ethical considerations. However, if human lymphoid tissues or organs are obtained in a surgical operation with informed consent and are implanted in NOG mice, the ethical problem may be solved. The advantages of the NOG mouse include the fact that it is an excellent recipient of human tissues of high frequencies and that human immune cells can remain in engrafted tissues for relatively long periods, so that organ implantation in NOG mice can overcome the disadvantage that may occur with implantation of human PBL as described previously.

Recently, we generated NOG mice with engraftment of adult human lymph nodes (LN) to promote human antigen-specific IgG production. In these experiments, one-quarter of the human LN, which were obtained with informed consent from patients with breast cancer, were implanted on the surface of the NOG spleen. These mice (abbreviated as LN-NOG) were immunized with DNP-KLH according to the protocol shown in Fig. 4. After immunization, the antibody level in the serum was examined every 2 weeks. Two out of 12 LN-NOG mice did not show any trace of engrafted LN tissues when they were sacrificed at the 7th week, and no T and B cells were detected in the spleen. In the remaining 10 LN-NOG mice bearing the engrafted LN, the antigen-specific antibody level was significant in 9 mice, with both IgM and IgG found in 7 mice and IgG alone and IgM alone in 1 mouse each (Fig. 5). One in 10 mice with engrafted LN died within 4 weeks but showed a slight increase of antigen-specific antibody in the 3rd week. In the LN-NOG spleen, numerous T and B cells were found and were analyzed by immunohistochemistry (Fig. 6) and flow cytometry (data not shown). In contrast, human-derived lymphoid cells were almost undetectable in murine LN of the inguinal and brachial regions, although decreased numbers of human T and B cells remained in the implanted human LN. These findings indicate that human immune cells of engrafted LN migrate into the contiguous spleen, where they proliferate and are activated efficiently

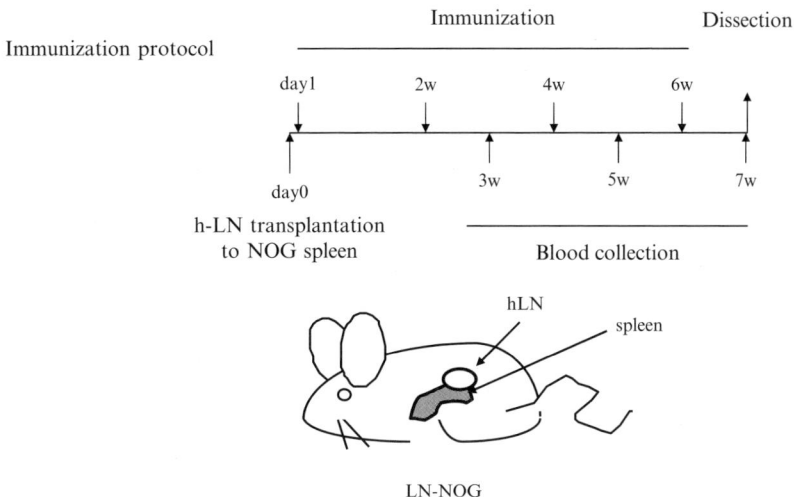

Fig. 4 Experimental protocol of lymph node engraftment and immunization in NOG mice. Nine-week-old NOG mice ($n=13$) were irradiated and transplanted with one-quarter of human lymph nodes on the spleen (LN-NOG). One day after transplantation, immunization with DNP-KLH was started biweekly. One week after each immunization, sera were collected for titration of antibodies. One week after the 3rd booster, LN-NOG mice were sacrificed for further analysis

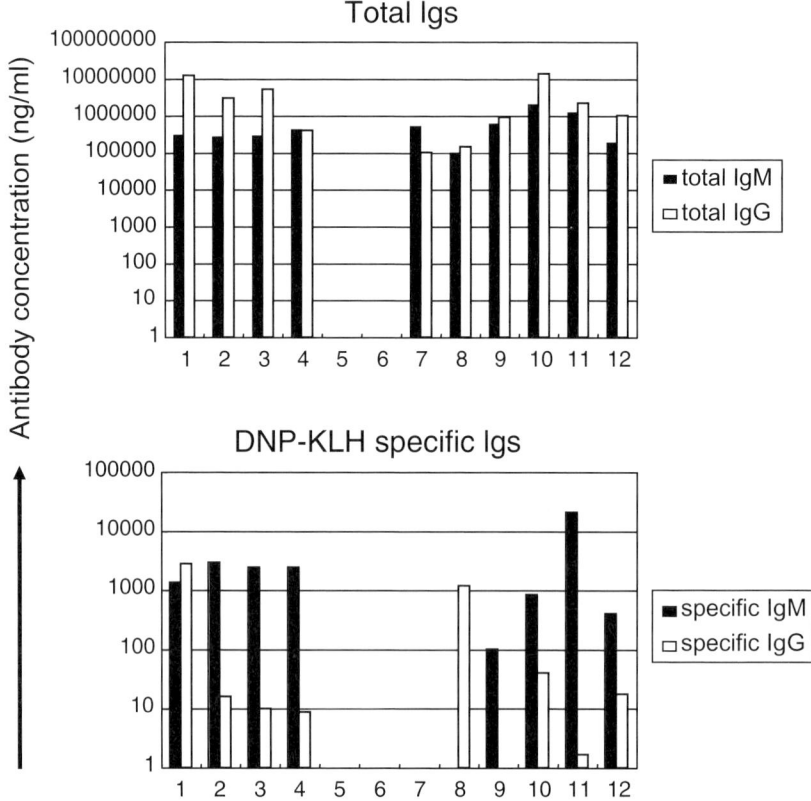

Fig. 5 Antibody production of antigen-specific IgG in LN-NOG mice immunized with DNP-KLH. NOG mice were immunized with DNP-KLH as described in Figs. 1 and 3. The serum concentration of antibodies (total, antigen specific IgG and IgM) was measured 3 times biweekly by ELISA. Each *bar* of the antibody concentration shows the highest level during immunization. *Closed bars*, human IgM. *Open bars*, human IgG. *Upper panel*, total human IgM and IgG. *Lower panel*, DNP-KLH-specific IgM and IgG. The *arabic numerals* on the *x-axis* represent the individual mouse numbers

with given antigens injected intraperitoneally. Our results are consistent with a previous report in which antigen-specific antibody was detected in SCID mice with engraftment of liver, thymus, skin fragments, and LN of the human embryo [3]. Taken together, it was suggested that if mature human cell components involved in immune reactions accumulate in a limited region for mutual interaction, suitable IgG antibody production for the given antigen can be induced in the mouse environment. In this context, LN-NOG mice are convenient for obtaining antigen-specific antibodies derived from human B cells, which would be risk-free therapeutic reagents.

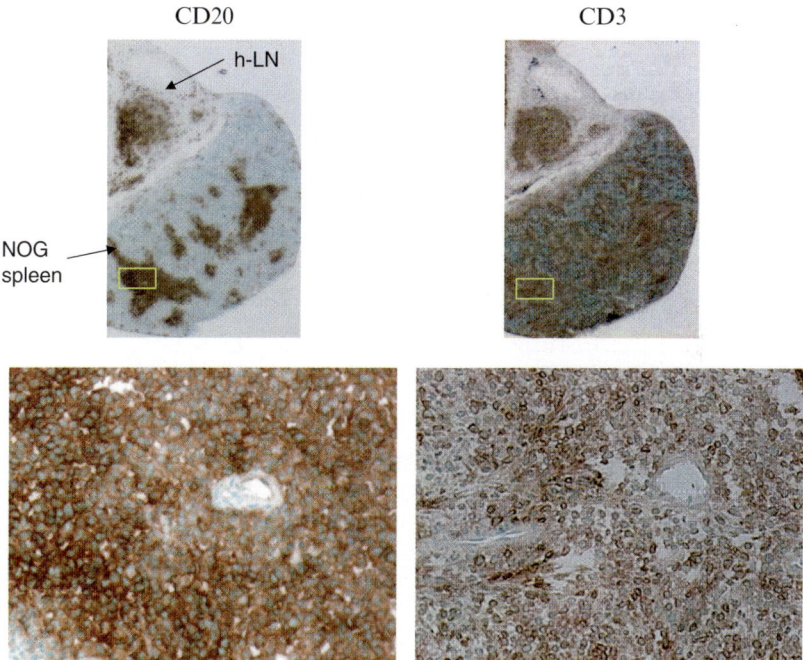

Fig. 6 Human T cells are detectable in NOG spleen and engrafted human lymph node. LN-NOG mouse tissues were subjected to immunohistochemical analysis by staining with anti-human CD20 or anti-human CD3 mAb. The *lower panels* show the magnification of the *boxed areas* of the *upper panels*

5 Conclusion

In NOG mice reconstituted with human CD34+CB cells, mature type human T and B cells developed well but antigen-specific IgG antibody was produced at an extremely low frequency and a low level, if at all, when these mice were immunized with an antigen. Human T cells developed in the mouse environment possess the potential to produce T cell-specific cytokines through TCR-mediated signaling, but it seems likely that they cannot efficiently recognize antigens presented by human APC and B cells because of human T cells positively selected by the murine MHC during the intrathymic development process in the mouse. As a result, memory T cells do not develop to help in class switching of immunoglobulin from IgM to IgG in B cells. Therefore, to obtain antigen-specific IgG antibody derived from human B cells, it is essential to overcome the restriction of the murine MHC in T cell development.

References

1. Abedi MR, Christensson B, Islan KB, Hammarstrom L and Smith CI (1992) Immunoglobulin production in severe combined immunodeficient (SCID) mice reconstituted with human peripheral blood mononuclear cells. Eur J Immunol 22:823-8
2. Anderson G, Owen JJ, Moore NC and Jenkinson EJ (1994) Thymic epithelial cells provide unique signals for positive selection of CD4+CD8+ thymocytes in vitro. J Exp Med 179:2027-31
3. Carballido JM, Namikawa R, Carballido-Perrig N, Antonenko S, Roncarolo MG and de Vries JE (2000) Generation of primary antigen-specific human T- and B-cell responses in immunocompetent SCID-hu mice. Nat Med 6:103-6
4. Glennie MJ and van de Winkel JG (2003) Renaissance of cancer therapeutic antibodies. Drug Discov Today 8:503-10
5. Hogan CJ, Shpall EJ, McNulty O, McNiece I, Dick JE, Shultz LD and Keller G (1997) Engraftment and development of human $CD34^+$-enriched cells from umbilical cord blood in NOD/LtSz-scid/scid mice. Blood 90:85-96
6. Ishikawa F, Yasukawa M, Lyons B, Yoshida S, Miyamoto T, Yoshimoto G, Watanabe T, Akashi K, Shultz LD and Harada M (2005) Development of functional human blood and immune systems in NOD/SCID/IL2 receptor gamma chain^{null} mice. Blood 106:1565-73
7. Ito M, Hiramatsu H, Kobayashi K, Suzue K, Kawahata M, Hioki K, Ueyama Y, Koyanagi Y, Sugamura K, Tsuji K, Heike T and Nakahata T (2002) NOD/SCID/γ_c^{null} mouse: an excellent recipient mouse model for engraftment of human cells. Blood 100:3175-82
8. Kametani Y, Shiina M, Katano I, Ito R, Ando K, Toyama K, Tsukamoto H, Matsumura T, Saito Y, Ishikawa D, Taki T, Ito M, Imai K, Tokuda Y, Kato S, Tamaoki N and Habu S (2006) Development of human-human hybridoma from anti-Her-2 peptide-producing B cells in immunized NOG mouse. Exp Hematol 34:1240-8
9. Kantor AB, Merrill CE, Herzenberg LA and Hillson JL (1997) An unbiased analysis of V_H-D-J_H sequences from B-1a, B-1b, and conventional B cells. J Immunol 158:1175-86
10. Kipps TJ (1989) The CD5 B cell. Adv Immunol 47:117-85
11. Li C, Ando K, Kametani Y, Oki M, Hagihara M, Shimamura K, Habu S, Kato S and Hotta T (2002) Reconstitution of functional human B lymphocytes in NOD/SCID mice engrafted with ex vivo expanded $CD34^+$ cord blood cells. Exp Hematol 30:1036-42
12. Markham RB and Donnenberg AD (1992) Effect of donor and recipient immunization protocols on primary and secondary human antibody responses in SCID mice reconstituted with human peripheral blood mononuclear cells. Infect Immun 60:2305-8
13. Matsumura T, Kametani Y, Ando K, Hirano Y, Katano I, Ito R, Shiina M, Tsukamoto H, Saito Y, Tokuda Y, Kato S, Ito M, Motoyoshi K and Habu S (2003) Functional CD5+B cells develop predominantly in the spleen of NOD/SCID/gamma-c^{null} (NOG) mice transplanted either with human umbilical cord blood, bone marrow, or mobilized peripheral blood CD34+ cells. Exp Hematol 31:789-97
14. McCune JM, Namikawa R, Kaneshima H Shults LD, Lieberman M and Weissman IL (1988) The SCID-hu mouse: murine model for the analysis of human hematolymphoid differentiation and function. Science 241:1632-9
15. Mirick G, Bradt B, Denardo S and Denardo G (2004) A review of human anti-globulin antibody (HAGA, HAMA, HACA, HAHA) responses to monoclonal antibodies. Not four letter words. Q J Nucl Med Mol Imaging 48:251-7
16. Mosier DE, Gulizia RJ, Baird SM and Wilson DB (1988) Transfer of a functional human immune system to mice with severe combined immunodeficiency. Nature 335:256-9
17. Namikawa R, Weilbaecher KN, Kaneshima H, Yee EJ and McCune JM (1990) Long-term human hematopoiesis in the SCID-hu mouse. J Exp Med 172:1055-1063
18. Nisitani S, Murakami M, Akamizu T, Okino T, Ohmori K, Mori T, Imamura M and Honjo T (1997) Preferential localization of human CD5+ B cells in the peritoneal cavity. Scand J Immunol 46:541-5

19. Osbourn J, Jermutus L and Duncan A (2003) Current methods for the generation of human antibodies for the treatment of autoimmune diseases. Drug Discov Today 8:845-51
20. Plum J, De Smedt M, Defresne MP, Leclercq G and Vandekerckhove B (1994) Human CD34+ fetal liver stem cells differentiate to T cells in a mouse thymic microenvironment. Blood 84:1587-93
21. Saito Y, Kametani Y, Hozumi K, Mochida N, Ando K, Ito M, Tokuda Y, Makuuchi H, Tajima T and Habu S (2002) The in vivo development of human T cells from CD34+ cells in the murine thymic environment. Int Immunol 14:1113-24
22. Sandhu J, Shpitz B, Gallinger S and Hozumi N (1994) Human primary immune response in SCID mice engrafted with human peripheral blood lymphocytes. J Immunol 152:3806-13
23. Sato T, Ohno S, Hayashi T, Sato C, Kohu K, Satake M and Habu S (2005) Dual functions of Runx proteins for reactivating CD8 and silencing CD4 at the commitment process into CD8 thymocytes. Immunity 22:317-28
24. Stern M and Herrmann R (2005) Overview of monoclonal antibodies in cancer therapy: present and promise. Crit Rev Oncol Hematol 54:11-29
25. Thomson BG, Robertson KA, Gowan D, Heilman D, Broxmeyer HE, Emanuel D, Kotylo P, Brahmi Z and Smith FO (2000) Analysis of engraftment, graft-versus-host disease, and immune recovery following unrelated donor cord blood transplantation. Blood 96:2703-11
26. Tomizuka K, Shinohara T, Yoshida H, Uejima H, Ohguma A, Tanaka S, Sato K, Oshimura M and Ishida I (2000) Double trans-chromosomic mice: maintenance of two individual human chromosome fragments containing Ig heavy and kappa loci and expression of fully human antibodies. Proc Natl Acad Sci USA 97:722-7
27. Yahata T, Ando K, Nakamura Y, Ueyama Y, Shimamura K, Tamaoki N, Kato S and Hotta T (2002) Functional human T lymphocyte development from cord blood CD34+ cells in non-obese diabetic/Shi-scid, IL-2 receptor gamma null mice. J Immunol 169:204-9
28. Yeoman H, Gress RE, Bare CV, Leary AG, Boyse EA, Bard J, Shultz LD, Harris DT and DeLuca D (1993) Human bone marrow and umbilical cord blood cells generate CD4+ and CD8+ single-positive T cells in murine fetal thymus organ culture. Proc Natl Acad Sci USA 90:10778-82

Humanized Immune System (HIS) Mice as a Tool to Study Human NK Cell Development

N. D. Huntington and J. P. Di Santo(✉)

1	Introduction	110
2	Human NK Cell Development	110
	2.1 In the Beginning...	111
	2.2 NK Cell Development in the Thymus	111
	2.3 Cytokine Dependence	114
	2.4 NK Cell Development in Lymph Nodes	114
	2.5 Immature NK Cells	115
	2.6 (Tran)Scripting the Fate of NK Cells	115
	2.7 Mature NK Cell Biology	116
3	Humanized Immune System Mice	116
	3.1 Balb/c Rag2$^{-/-}\gamma_c^{-/-}$ HIS Mice	117
	3.2 Development of Human NK Cells in HIS Mice	118
	3.3 Considerations and Uses of HIS Mice	119
4	Summary	120
References		120

Abstract The study of human hematopoiesis is conditioned by access to nondiseased human tissue samples that harbor the cellular substrates for this developmental process. Technical and ethical concerns limit the availability to tissues derived from the fetal and newborn periods, while adult samples are generally restricted to peripheral blood. Access to a small animal model that faithfully recapitulates the process of human hematopoiesis would provide an important tool. Natural killer (NK) cells comprise between 10% and 15% of human peripheral blood lymphocytes and appear conserved in several species. NK cells are implicated in the recognition of pathogen-infected cells and in the clearance of certain tumor cells. In this chapter, we discuss NK cell developmental pathways and the use of humanized murine models for the study of human hematopoiesis and, in particular, human NK cell development.

J. P. Di Santo
Cytokine and Lymphoid Development Unit, Immunology Department, Institut Pasteur,
25 rue du Docteur Roux, Paris 75724, France
disanto@pasteur.fr

Abbreviations BM: bone marrow; DC: dendritic cell; FcγRIII: Fc gamma receptor III; HIS: humanized immune system; HLA: human leukocyte antigen; HSC: hematopoietic stem cell; IL: interleukin; KIR: killer inhibitory receptor; LN: lymph node; MHC: major histocompatibility complex; N-CAM: neural cell adhesion molecule; NK: natural killer; TAP-1: transporters associated with antigen processing 1

1 Introduction

Since experimentation in both humans and nonhuman primates is limited by factors such as ethics, expense, facilities, and material, there has long been a great interest in generating accurate in vivo small animal models to study human immunology. For almost 20 years, researchers have been investigating the possibility of transplanting and studying human tissue in mice. Humanized immune system mice (HIS mice) that are capable of developing and maintaining functional human immune cells could prove a valuable tool for studying human immune disease and for development of vaccines and therapeutics in an accessible, valid, and controlled environment (Legrand et al. 2006b; Macchiarini et al. 2005). While studies in HIS mice will likely never be as accurate or informative as controlled human trials, they should allow substantial and valuable preclinical information. In addition, HIS mice could allow a thorough in vivo analysis of human lymphocyte development by providing a source of human lymphocytes from tissues that are rarely available or accessible for study in humans.

Our laboratory has a long-standing interest in the biology of natural killer (NK) cells. Recent reviews on lymphocyte development have underlined the fact that our understanding of NK cell development lags far behind that of B and T cells (Di Santo 2006). Even more apparent is the knowledge gap between what is known about murine NK cell development and that of humans. As such, we have very little knowledge of the phenotype, differentiation, function, and tissue distribution of immature NK cells or NK cell precursors in humans (reviewed in Blom 2006).

2 Human NK Cell Development

NK cells act as important contributors to innate immune responses by recognizing transformed or stressed cells and responding rapidly with the release of cytolytic granules and cytokines (Diefenbach et al. 2001; Roder et al. 1979; Biron et al. 1999; Trinchieri 1989). While mature NK cells can be found in numerous locations such as spleen, blood, lymph node (LN), lung, liver, peritoneal cavity, thymus, and bone marrow (BM), the identity of NK cell precursors, their location, and transcription factors involved in NK cell development are not well defined.

2.1 In the Beginning...

Human fetal NK cell precursors have been described in the liver, thymus, BM, and LN, with most of these studies identifying different cell populations within the CD34+ hematopoietic stem cell (HSC) compartment that can generate mature NK cells in the presence of stromal cells and certain cytokines (Blom and Spits 2006; Spits et al. 1995). NK cells expressing the markers CD56, CD16, and CD94 can be derived from both CD34+CD38+ and CD34+CD38− fetal liver cells when cultured with IL-7, Flt-3L, and IL-15 (Jaleco et al. 1997). CD34+CD38+ cells appeared to be more restricted to the NK cell lineage as they failed to develop into T cells, in contrast with the CD34+CD38− population.

Early work reported that CD7 expression within the adult BM CD34+ HSC population identified a NK cell precursor population with enhanced cloning efficiency (Miller et al. 1994). In this study, NK cell precursor activity was correlated with the level of CD7 expression, with BM-CD34+CD7^{hi} cells possessing higher cloning efficiency than CD34+CD7^{dim} cells. A related precursor population bearing the CD34^{hi}CD10+Lin−CD45RA+CD38+c-kit− phenotype gave rise to B, NK, and dendritic cells (DC) (Galy et al. 1995). More recently, the expression of CD10 and CD7 within the CD34+Lin− BM population was used to demonstrate that cells differentially expressing CD7 versus CD10 had NK and B cell precursor activity, respectively (Rossi et al. 2003); a similar observation was also reported with umbilical cord blood CD34+ cells (Haddad et al. 2004). Expression of CD127 (IL-7Rα chain) was also observed on a fraction of BM CD34+CD10+ cells that were also CD45RA+CD38+CD24−c-kit^{dim}, and while these cells were able to differentiate into NK cells, they appeared to be more restricted to the B cell lineage (Ryan et al. 1997).

2.2 NK Cell Development in the Thymus

While it is well appreciated that the BM is the major location of NK cell development in adults, we recently identified a novel thymic pathway of NK cell development in mice, which may also exist in humans (Di Santo and Vosshenrich 2006; Vosshenrich et al. 2006). We found that thymic NK cells were phenotypically distinct from most NK cells found in the spleen, blood, and liver, bearing high levels of CD127 at their cell surface. CD127+ NK cells were functionally distinct from those found in the spleen as they possessed less cytolytic activity and intracellular granzyme B, but contained higher amounts of intracellular IFN-γ, TNF-α, and GM-CSF when restimulated in vitro (Vosshenrich et al. 2006). Interestingly, besides the thymus, CD127+ NK cells were also found in the LN, where they represented about one-quarter of resident NK cells. From studying thymic grafts and athymic mice, we confirmed that the thymus exported CD127+ NK cells to the LN and was necessary for their development (Vosshenrich et al. 2006). These data imply that precursor cells that seed the thymus undergo a different developmental program from that of those residing in the BM, although the phenotype of the NK precursors in the thymus and their relationship to BM NK precursors remains undefined.

Concerning the factors that regulate the thymic developmental pathway, we demonstrated that thymic CD127$^+$ NK development required IL-7 and the Zinc-finger transcription factor GATA-3. While Rag2$^{-/-}$IL-7$^{-/-}$ mice contained thymic NK cells, the majority of these cells lacked CD127 expression, which directly contrasted with Rag2$^{-/-}$ thymic NK cells. Furthermore, Rag2$^{-/-}$ mice reconstituted with fetal liver lacking GATA-3 also failed to develop thymic CD127$^+$ NK cells (Vosshenrich et al. 2006). GATA-3 is of particular interest in this pathway, because we have previously shown that GATA-3 is not essential for BM-dependent NK cell development as NK cells are clearly present in the BM and spleen of mice reconstituted with GATA-3-null fetal liver cells, although they do appear less mature than those in control mice (Samson et al. 2003). GATA-3 is also required for normal NK cell homeostasis in the liver and for optimal IFN-γ production from splenic and hepatic NK cells; however, the mechanism behind these phenotypes is still unclear (Samson et al. 2003).

Given the hallmarks of thymic-derived NK cells in the mouse, we asked whether a subset of human NK cells possessed similar properties. Indeed, we observed CD127 expression on human thymic NK cells and also on a subpopulation of peripheral blood NK cells (CD56^{hi}CD16$^-$). Previous reports showed that CD56^{hi}CD16$^-$ NK cells in lymph node have variable expression of CD127, whereas CD56^{dim}CD16$^+$ NK cells were CD127 negative (Freud and Caligiuri 2006; Hanna et al. 2004), suggesting that the human and mouse NK cell subsets may subserve similar functions. Supporting this hypothesis, murine CD127$^+$ NK cells and human peripheral blood CD56^{hi}CD16$^-$ NK cells are better producers of cytokines but less cytotoxic than their CD127$^-$ counterparts (Vosshenrich et al. 2006). It is tempting to speculate, given these data, that a thymic pathway of NK cell development may also exist in humans (Fig. 1), but more data are needed to confirm this.

The human thymus contains a minor population of CD34$^+$CD38^{dim} human thymocytes that are capable of differentiating into NK cells in the presence of stem cell factor (SCF), IL-7, and IL-2 (Res et al. 1996). In addition, when fetal human thymocytes (16-22 gestational weeks) were cultured in IL-7 and IL-15 on methylcellulose for 2 weeks, colonies of CD56$^+$CD3$^-$ cells were generated that were able to lyse K562 cells (Sato et al. 1999). These colonies of NK cells were enriched when Lin$^-$CD34^{hi} fetal thymocytes were cultured in the same manner, whereas Lin$^-$CD34$^+$ and Lin$^-$CD34$^-$ fetal thymocytes displayed less NK cell precursor activity (Sato et al. 1999). With fetal-based assays for conditions favoring T cell differentiation, both T cells and NK cells were derived from both CD34$^+$CD1$^-$ and CD34$^+$CD1$^+$ thymocyte precursors (Sarun et al. 1998), whereas others have shown that the acquisition of CD1a on hematopoietic progenitors in the human thymus correlates with a loss of capacity to develop into NK cells and the commitment to the T cell lineage (Spits et al., 1998). The growth of human fetal thymocytes in vitro appears to be best achieved when a combination of IL-15, IL-7, and c-kit ligand (c-kitL) are used; however, as a single factor, only IL-15 supported NK cell colony formation

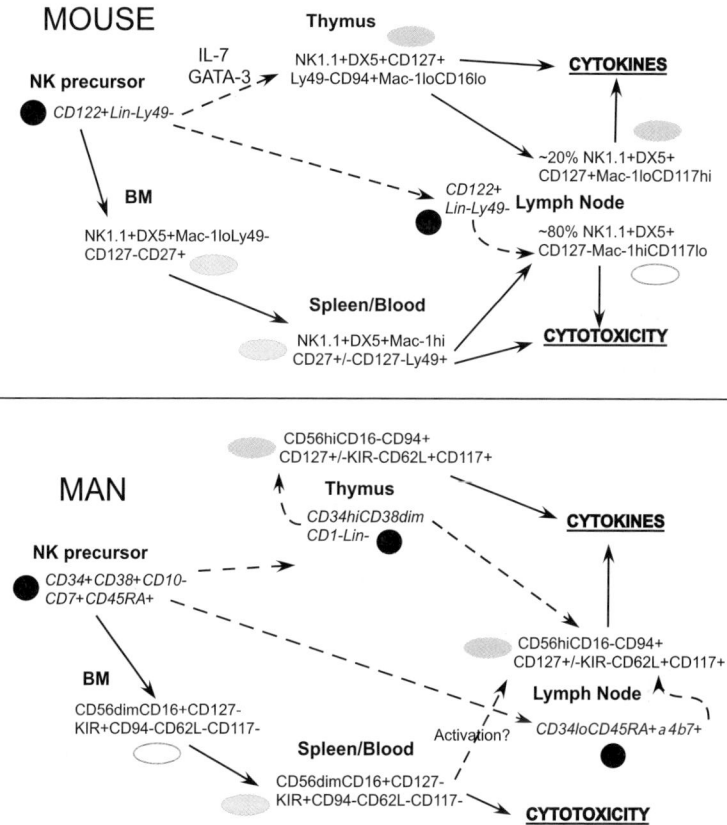

Fig. 1 Potential overlapping pathways for NK development in human and mouse. In mouse, CD122+Lin− NK cell precursors are capable of seeding the BM, where they develops into immature NK1.1+DX5+Mac-1loLy49−CD127− NK cells that mature into NK1.1+DX5+Mac-1hiLy49+CD127− NK cells that frequent the spleen and blood. This CD122+Lin− NK cell precursor might also seed the thymus, giving rise to a unique NK1.1+DX5+Mac-1loLy49−CD127+ population that is dependent on IL-7 and GATA-3. This CD127+ NK cell subset is also found in lymph nodes and appears more efficient at producing cytokines but less cytotoxic compared to the CD127− NK cell subset. Similar pathways may exist in humans, with a potential CD34+CD38+CD7+CD45RA+ NK cell precursor being responsible for BM-dependent NK cell development, which accounts for CD56dimCD16+KIR+CD127− mature NK cells in the blood and spleen. This NK cell precursor might also seed the thymus or lymph node, where a different development program may result in the CD56hiCD16−KIR−CD127+/− NK cell population that frequents these organs. As in mice, differential functions between NK cell subsets exist in humans, with the hypothetical thymus-derived NK cell (CD56hiCD16−KIR−CD127+/−) having better cytokine production but worse cytotoxicity than BM-derived (CD56dimCD16+KIR+CD127−) NK cells, which is the actual case in mice. Alternatively, CD56hiCD16−KIR−CD127+/− NK cells may represent an activated form of CD56dimCD16+KIR+CD127− NK cells

(Sato et al. 1999). These data indicate that the thymus has the potential to contribute to NK cell development in humans; however, whether this actually occurs in vivo and the physiological relevance of this pathway are unknown.

2.3 Cytokine Dependence

Cytokines play a critical role in the development of human NK cells, as evidenced by humans with mutations in the gene encoding the common gamma chain (γ_c) that display dramatically impaired NK cell development (Buckley 2004). Several γ_c-dependent cytokines (including IL-2, IL-7, IL-15, and IL-21) may contribute to NK cell phenotype in these patients (Buckley 2004). A proline to serine substitution (P132S) in the CD127 gene has also been identified in three patients presenting a severe combined immunodeficiency (SCID) phenotype (Roifman et al. 2000). This mutation severely compromised the affinity of CD127 to bind IL-7, resulting in impaired signal transduction and an obvious reduction in peripheral T cells. However, unlike mutations in γ_c, NK cell numbers and function appeared unaffected (Roifman et al. 2000). Nevertheless, the study of NK cells in humans with CD127 mutations was restricted to the expression of CD56 alone and did not include differential analysis of the peripheral NK cell subsets (Roifman et al. 2000). Thus it remains possible that IL-7 may be required in humans for the development of the $CD56^{hi}CD16$ NK cells, analogous to what has been described in mice (Vosshenrich et al. 2006).

The ability of c-kitL to synergize with IL-15 and IL-7 in enhancing NK cell development from $CD34^+$ cells may involve distinct mechanisms (proliferation of stem cells, survival of developing NK cells, etc.). Indeed, it has been shown that while c-kitL alone does not promote proliferation of $CD56^{hi}$c-kit$^+$ NK cells, it does increase their Bcl-2 expression and prevent apoptosis in the absence of growth factors (Carson et al. 1994). Thus c-kitL may enhance the expansion of NK cells derived from $CD34^+$ precursors in the presence of IL-15 (Mrozek et al. 1996). In contrast, factors derived from thymic epithelial cells (TEC) and Notch signaling appear detrimental to NK cell development. Notch signaling in human progenitor cells is thought to be critical in directing them to the T cell lineage and impairing NK cell commitment (De Smedt et al. 2005), whereas the presence of TEC-derived soluble factors inhibited NK cell development and differentiation from $CD34^+$ thymocytes in the presence of IL-15 (Le et al. 2001).

2.4 NK Cell Development in Lymph Nodes

In addition to the BM and possibly the thymus, the LN has been recently proposed as a site of human NK cell development since it was shown to harbor NK cell precursors. A $CD34^{lo}CD45RA^{hi}\alpha 4\beta 7^+$ cell population was observed at low levels among adult BM and peripheral blood $CD34^+$ cells (<1% and ~6%, respectively), but represented over 95% of LN $CD34^+$ cells (Freud et al. 2005). The $CD34^{lo}CD45RA^{hi}\alpha 4\beta 7^+$ LN subset resides in parafollicular T cell regions and develops into $CD56^{hi}CD16^-$ NK cells at high efficiency in the presence of IL-15 or IL-2 in vitro (Freud et al. 2005). While it is not clear what other cell lineages may be derived from LN $CD34^{lo}CD45RA^{hi}\alpha 4\beta 7^+$ cells, the peripheral location of $CD34^{lo}CD45RA^{hi}\alpha 4\beta 7^+$ cells indicates that they are likely to be committed hematopoietic precursors. One possibility is that LN-resident CD34^{lo}CD45RA$^+$α4β7^{hi}CD7$^+$ NK precursors are derived from the CD34$^+$CD45RA$^+$CD7$^+$CD10$^+$ or CD34$^+$CD45RA$^+$CD7$^+$CD10$^-$ BM lymphocyte precursors that possess NK cell precursor activity (Galy et al. 1995; Miller et al. 1994; Rossi et al. 2003). It would be interesting to know whether an equivalent population exists in the LN of mice and whether these NK cell precursors are derived directly from the BM or supplied by a thymic route as shown for CD127$^+$ NK cells.

2.5 Immature NK Cells

Evidence for discrete stages of NK cell commitment and immature NK cell phenotype and function in humans is limited. Early studies demonstrated that a population of noncytotoxic CD56$^-$NKR$^-$P1A$^+$ cells are generated in the presence of IL-2 and a BM stromal cell line expressing human stem cell factor (Bennett et al. 1996). This cell population has the characteristics of an immature NK cell and acquired a mature phenotype (CD56$^+$, cytotoxicity, and IFN-γ production) after a further 10 days of culture in the presence of IL-2 and IL-12 (Bennett et al. 1996). Murine NK cells express inhibitory receptors for MHC I, such as Ly49 family members and CD94/NKG2A, which are acquired during immature stages of development in the BM (Di Santo 2006; Kim et al. 2002; Yokoyama et al. 2004). When human umbilical cord blood CD34$^+$Lin$^-$CD38$^-$ cells are cultured on a murine fetal liver line with IL-15, they acquire equivalent human inhibitory receptors for MHC I, notably members of the killer immunoglobulin-like receptor (KIR) family and CD94 (Miller and McCullar 2001). The presence of either IL-15 or IL-2 and contact with the fetal liver line was essential for expression of these receptors on differentiating NK cells. Furthermore, analysis of single CD34$^+$Lin$^-$CD38$^-$ cells cultured in IL-15 or IL-2 suggests that the pattern of KIR expression is determined after this progenitor stage, as the NK cells that developed from one precursor were polyclonal in respect to KIR expression (Miller and McCullar 2001).

2.6 (Tran)Scripting the Fate of NK Cells

Transcription factors play a critical role in dictating the lineage fate of HSC. From studies in mice, we know that transcription factors including GATA-3, Ets-1, IRF-1/2, and T-bet are involved in NK cell differentiation, as mice lacking these factors demonstrate abnormal NK cell phenotypes in terms of cell number, maturation, or function (Barton et al. 1998; Di Santo 2006; Lohoff et al. 2000; Ogasawara et al. 1998; Samson et al. 2003; Townsend et al. 2004). Similarly, the helix-loop-helix inhibitor Id2 is also required for NK cell development in mice (Ikawa et al. 2001). Indeed, overexpression of Id2 and Id3 in human CD34$^+$ cells by retrovirus-mediated gene transfer promoted NK cell development in fetal thymic organ culture (FTOC) while blocking the development into T cells and B cells (Heemskerk et al. 1997).

The same group also demonstrated that the transcription factor Spi-B, which is expressed in CD34$^+$ cells and promotes plasmacytoid DC development, inhibits the development of human NK cells in a FTOC system (Schotte et al. 2003).

2.7 Mature NK Cell Biology

In contrast to immature NK cells and precursors, there is a reasonable level of knowledge concerning the biology of mature human NK cells. The existence of two subsets of peripheral human NK cells is well documented and has been the focus of numerous recent reviews (Farag and Caligiuri 2006; Freud and Caligiuri 2006). The distinguishing feature of these two human NK cell subsets is their differential surface expression of CD16 (FcγRIII) and CD56 (N-CAM) (Lanier et al. 1986, 1989). The CD56^{dim}CD16$^+$ represents around 90% of human peripheral blood NK cells, while the remaining NK cells are CD56^{hi}CD16$^-$. The CD56^{hi}CD16$^-$ NK cell subset express high levels of CD94/NKG2A and are CCR7$^+$CD62L$^+$CD25$^+$c-kit$^+$ but largely lack expression of KIRs. In contrast, the CD56^{dim}CD16$^+$ NK cells are mostly negative for CD94/NKG2A and are CCR7$^-$CD62L$^-$CD25$^-$c-kit$^-$ but express high levels of KIRs. These differential surface phenotypes equate to vastly different functional capacities. CD56^{hi}CD16$^-$ NK cells produce greater amounts of IFN-γ, TNF-β, GM-CSF, TNF-α, IL-13, and IL-10 compared to CD56^{dim}CD16$^+$ NK cells, and they also respond to lower concentrations of IL-2. On the other hand, CD56^{dim}CD16$^+$ NK cells have a greater cytotoxic capacity (Cooper et al. 2001; Farag and Caligiuri 2006; Freud and Caligiuri 2006). A major question remains unresolved concerning the relationship between CD56^{dim}CD16$^+$ and CD56^{hi}CD16$^-$ NK cells. While some favor the idea that CD56^{hi}CD16$^-$ NK cells are precursors of CD56^{dim}CD16$^+$ NK cells (Freud and Caligiuri 2006), others propose that CD56^{hi} NK cells represent activated or differentiated CD56^{dim} NK cells, as a CD56^{hi} phenotype can be achieved by culturing CD56^{dim} NK cells in IL-12 (Loza and Perussia 2004).

Interestingly, the ratio of CD56^{dim}CD16$^+$ to CD56^{hi}CD16$^-$ NK cells in the LN contrasts with that of the peripheral blood, with the majority of the NK cells being CD56^{hi}CD16$^-$. Since CD56^{hi}CD16$^-$ NK cells predominantly secrete cytokines including IFN-γ after activation, it has been suggested that as in mice (Martin-Fontecha et al. 2004), human LN NK cells can help to prime Th1 cells (Morandi et al. 2006).

3 Humanized Immune System Mice

Various murine strains harboring genetic mutations that affect their capacity to reject xenografts are capable of receiving and facilitating the development of transplanted human hematopoietic cells. As some of these models are discussed in detail in other chapters in this volume, we focus our attention on the Rag2$^{-/-}$γ$_c^{-/-}$ model.

Early experimentation aimed at overcoming the mouse-human transplantation barrier involved grafting of human fetal liver, fetal LN, and fetal thymic tissue into immunodeficient mouse strains (lacking B and T cells) such as C.B-17 SCID and NOD/SCID (McCune 1997; McCune et al. 1988). In the resulting SCID-hu mice, single positive human T cells and human IgG were observed in circulation. Human cells in SCID-hu mice could also be productively infected with HIV-1 (McCune et al. 1988; Namikawa et al. 1988). These models were improved by including mutations and/or conditioning regimes to reduce host (murine) NK cell numbers or function, as NK cells appeared to be a limiting factor in the success of earlier models (Legrand et al. 2006b; Macchiarini et al. 2005).

3.1 Balb/c Rag2$^{-/-}\gamma_c^{-/-}$ HIS Mice

Arguably the most successful example of humanized immune system (HIS) mice is the work of Traggiai et al., who transplanted CD34$^+$ umbilical cord blood cells in newborn Balb/c Rag2$^{-/-}\gamma_c^{-/-}$ mice via the intrahepatic (IH) route (Traggiai et al. 2004). This group reasoned that "since the liver contributes to perinatal hematopoiesis, and the hemato-lymphoid system expands most significantly during the first weeks of life... human hematopoietic stem and progenitor cells transplanted into the liver of immunodeficient newborn mice might find better conditions to engraft, expand, and reconstitute a human immune system." This procedure was extremely effective, with human hematopoietic cells being detected in all engrafted mice, and clear increases in lymphoid tissue cellularity were observed (Traggiai et al. 2004). This model appears practical for the study of human immune cells, as not only is engraftment effective but the resulting human lymphocytes appear functional in certain experimental settings. T cell areas and B cell follicles (including occasional germinal centers) were observed in the spleen, and after immunization with tetanus toxoid (TT) anti-tetanus IgG could be elicited (Traggiai et al. 2004).

The HIS model is not restricted to cord blood engraftment or engraftment into the liver, as newborn Balb/c Rag2$^{-/-}\gamma_c^{-/-}$ mice receiving human fetal liver CD34$^+$ cells via the intraperitoneal (IP) route achieve similar levels of human hematopoietic cell reconstitution (Gimeno et al. 2004; Legrand et al. 2006a). The level of human chimerism is comparable when CD34$^+$ fetal liver cells are injected either IH or IP (Legrand, Weijer, and Spits, unpublished).

Recently, the Balb/c Rag2$^{-/-}\gamma_c^{-/-}$ HIS mouse was used to study human disease, particularly infection by HIV (Baenziger et al. 2006; Gorantla et al. 2006; Zhang et al. 2006), opening the road to developing new therapeutic approaches. However, HIS mice also provide an in vivo tool allowing for a more in-depth analysis of human lymphocyte development by providing access to human lymphocytes from tissues that are typically difficult to obtain. Along these lines, our laboratory is currently investigating the use of Balb/c Rag2$^{-/-}\gamma_c^{-/-}$ HIS mice to study human NK cell development and function.

The protocol adopted in our laboratory for the generation of Balb/c Rag2$^{-/-}\gamma_c^{-/-}$ HIS mice is essentially identical to that described by the Spits group (Gimeno et al. 2004; Legrand et al. 2006a). Human fetal liver (14-20 weeks of gestation) is obtained, and CD34$^+$ human fetal liver cells are purified by magnetic separation and/or further sorted according to cell surface phenotype. Newborn Balb/c Rag2$^{-/-}\gamma_c^{-/-}$ mice (typically 3-5 days old) are sublethally irradiated (330 rads) and injected IH with 0.5 x 10^6 CD34$^+$ cells. Reconstituted mice are housed in micro-isolator units, and all procedures are performed under laminar flow. Peripheral blood is collected at 4 weeks after transplantation and analyzed for human CD45 expression by FACS to determine engraftment efficiency.

3.2 Development of Human NK Cells in HIS Mice

Both human CD56^{hi}CD16$^-$ and CD56^{dim}CD16$^+$ NK cells develop in Balb/c Rag2$^{-/-}\gamma_c^{-/-}$ HIS mice, although they represent a relatively low proportion in the BM, thymus, and spleen (Gimeno et al. 2004 and our unpublished data). We observe that human NK cell in Balb/c Rag2$^{-/-}\gamma_c^{-/-}$ HIS mice express most of the cell surface antigens found on fresh human NK cells (including CD94, NKp30, NKp46, CD11b, CD62L, inhibitory KIRs, etc.) and on activation accumulate intracellular IFN-γ and granzyme-B (NDH and JPD, manuscript in preparation). Thus, human NK cells that develop in HIS mice resemble normal human NK cells, suggesting that HIS mice may provide a new tool to study human NK cell development and function.

The expression of CD94 and members of the KIR family has some important implications for NK cell function in this model. Like murine NK cells, human NK cells express inhibitory receptors for MHC class I molecules. The presence of these receptors forms the basis on the "missing self hypothesis," which states that NK cells are "educated" to scrutinize target cell MHC expression and only target cells lacking the expression of MHC I will evoke a response from host NK cells (Karre et al. 1986). However, evidence from both MHC I-deficient mice and humans and also NK cells lacking expression of receptors for MHC I suggested that NK cell inhibitory receptors and their MHC ligands may play an important role during the functional differentiation of NK cells as well (Anfossi et al. 2006; Bix et al. 1991; Furukawa et al. 1999; Kim et al. 2005). NK cells from a TAP-1-deficient patient showed no cytotoxicity against target cells lacking MHC I molecules despite the presence of class I-recognizing NK receptors (Furukawa et al. 1999). In a similar fashion, NK cells from MHC I-deficient mice fail to be activated when triggered through activating receptors (Fernandez et al. 2005; Kim et al. 2005). These observations form the foundations of "licensing" or "education" models for achieving NK cell tolerance, where an inhibitory receptor/MHC I interaction will either activate directly or modify other signals to developing NK cells, allowing them to gain full functionality (Anfossi et al. 2006; Bix et al. 1991; Furukawa et al. 1999; Kim et al. 2005; Yokoyama and Kim 2006). Recently, a population of MHC I inhibitory receptor-negative NK cells was reported in both mice (Fernandez et al. 2005; Kim et al. 2005) and humans (Anfossi et al. 2006). While human KIRs⁻NKG2A⁻ NK cells are distributed evenly between the CD56^{hi}CD16⁻ and CD56^{dim}CD16⁺ NK subsets, they are less sensitive to activation by cytokines or activating receptors compared to the KIR⁺NKG2A⁺ NK cells as determined by induction of CD107a and IFN-γ expression (Anfossi et al. 2006). If NK cells require interactions with MHC class I molecules to be fully functional in Balb/c Rag2$^{-/-}$γ$_c$$^{-/-}$ HIS mice, then the cognate MHC I ligands may need to be expressed in these mice during NK cell development.

The human NK cell receptors for ligands encoded by class I genes include CD94/NKG2A/C/E receptors recognizing HLA-E, KIR2DL4 binding to HLA-G, and other KIRs recognizing HLA-A, HLA-B, and HLA-C (Parham 2006). Population studies of KIR gene expression has revealed that individuals expressing a KIR member along with its cognate ligand had a greater percentage of NK cells expressing that particular KIR receptor compared to those with the same KIR gene but lacking the ligand (Parham 2006). If this is so, then the functionality of NK cells that develop in Balb/c Rag2$^{-/-}$γ$_c$$^{-/-}$ HIS mice may be conditioned by cross-reactivity with mouse MHC I or by selection on MHC I ligands expressed by developing human hematopoietic cells. Furthermore, the HLA and KIR genes expressed by each donor may also influence the expression of inhibitory MHC I receptors on human NK cells generated in Balb/c Rag2$^{-/-}$γ$_c$$^{-/-}$ HIS mice.

3.3 Considerations and Uses of HIS Mice

Another important factor to consider for human NK cell development in Balb/c Rag2$^{-/-}$γ$_c$$^{-/-}$ HIS mice is the availability of the relevant cytokines that are required for NK cell generation and survival. Two sources are available: human cytokines secreted during lymphocyte development and mouse cytokines that may cross-react with human cells. The pleiotropic cytokine IL-15 appears the most critical in this process, as mice that are deficient in IL-15, IL-15 receptor (IL-15R) subunits (α, β, and γ), or IL-15R signaling molecule Jak3 lack peripheral NK cells (DiSanto et al. 1995; Kennedy et al. 2000; Lodolce et al. 1998; Suzuki et al. 1997; (Nosaka et al. 1995; Park et al. 1995). The response of human NK cells to murine IL-15 is not known. Whether donor-derived human cells expressing human IL-15 bound to IL-15Rα (Koka et al. 2003) drive NK cell development in Balb/c Rag2$^{-/-}$γ$_c$$^{-/-}$ HIS mice is equally unknown. Given our recent finding for a requirement for IL-7 in the development of thymus-derived NK cells in mice (Vosshenrich et al. 2006), the bioavailability of human IL-7 during development may also be required to reveal an equivalent human subset in Balb/c Rag2$^{-/-}$γ$_c$$^{-/-}$ HIS mice.

While Balb/c Rag2$^{-/-}$γ$_c$$^{-/-}$ HIS mice represent a potentially extremely powerful tool in investigating the biology of the human immune system, there are numerous limitations and areas for improvement in this model. Understanding factors such as the variability in reconstitution, unequal development of human lymphoid cells (i.e., B cells>T cells>NK cells), and stability of lymphocyte homeostasis is central to improving the Balb/c Rag2$^{-/-}\gamma_c^{-/-}$ HIS model. Furthermore, as Balb/c Rag2$^{-/-}\gamma_c^{-/-}$ mice lack lymphoid tissue inducer cells (which drive the formation of most peripheral LNs), the absence of LN may affect NK cell homeostasis at multiple levels (development, survival) and could also influence immune responses (initiation and maintenance of immune responses).

Nevertheless, the HIS model offers an unique opportunity to investigate human hematopoiesis in general and human NK cell differentiation in particular. For example, Balb/c Rag2$^{-/-}\gamma_c^{-/-}$ HIS mice provide the means to directly regulate the expression of transcription factors, antiapoptotic factors, cell surface receptors, and signaling proteins during human lymphocyte development by introducing short hairpin RNA or viral overexpression constructs into the CD34$^+$ human stem cells before engraftment (Gimeno et al. 2004; Schotte et al. 2004). In addition, the genetics of the hosts can be further improved through transgenic expression of human cytokines, hormones, HLA molecules. and other receptors in the murine host. These improvements in the Balb/c Rag2$^{-/-}\gamma_c^{-/-}$ HIS model should create the conditions suitable for preclinical testing of novel therapeutic agents in the combat against human disease.

4 Summary

Humanized mouse strains provide useful tools for the study of human lymphocyte development, for understanding the pathophysiology of human disease, and for development of novel therapeutics. The Balb/c Rag2$^{-/-}\gamma_c^{-/-}$ HIS model is efficiently engrafted with human cells and develops functional lymphocyte populations. By using the Balb/c Rag2$^{-/-}\gamma_c^{-/-}$ HIS model, it is possible to address novel questions concerning human NK cell development and function. Furthermore, since NK cells are important in viral immune responses (Lodoen and Lanier 2005), fundamental knowledge of NK cell biology gained by using the Balb/c Rag2$^{-/-}\gamma_c^{-/-}$ HIS model may impact on our understanding of human immune responses and help in designing more effective antiviral vaccines.

Acknowledgements This work is supported by grants from Institut Pasteur, INSERM, and Ligue National Contre le Cancer and a Grand Challenges in Global Health grant from the Bill & Melinda Gates Foundation. We would like to thank Erwan Corcuff, Nicolas Legrand, Hergen Spits, and Kees Weijer for their assistance and collaboration.

References

Anfossi, N., Andre, P., Guia, S., Falk, C. S., Roetynck, S., Stewart, C. A., Breso, V., Frassati, C., Reviron, D., Middleton, D. et al. (2006). Human NK cell education by inhibitory receptors for MHC class I. Immunity 25, 331-342.

Baenziger, S., Tussiwand, R., Schlaepfer, E., Mazzucchelli, L., Heikenwalder, M., Kurrer, M. O., Behnke, S., Frey, J., Oxenius, A., Joller, H. et al. (2006). Disseminated and sustained HIV infection in CD34+ cord blood cell-transplanted Rag2$-/-\gamma_c-/-$ mice. Proc Natl Acad Sci USA *103*, 15951-15956.

Barton, K., Muthusamy, N., Fischer, C., Ting, C. N., Walunas, T. L., Lanier, L. L., and Leiden, J. M. (1998). The Ets-1 transcription factor is required for the development of natural killer cells in mice. Immunity *9*, 555-563.

Bennett, I. M., Zatsepina, O., Zamai, L., Azzoni, L., Mikheeva, T., and Perussia, B. (1996). Definition of a natural killer NKR–P1A+/CD56–/CD16– functionally immature human NK cell subset that differentiates in vitro in the presence of interleukin 12. J Exp Med *184*, 1845-1856.

Biron, C. A., Nguyen, K. B., Pien, G. C., Cousens, L. P., and Salazar-Mather, T. P. (1999). Natural killer cells in antiviral defense: function and regulation by innate cytokines. Annu Rev Immunol *17*, 189-220.

Bix, M., Liao, N. S., Zijlstra, M., Loring, J., Jaenisch, R., and Raulet, D. (1991). Rejection of class I MHC-deficient haemopoietic cells by irradiated MHC-matched mice. Nature *349*, 329-331.

Blom, B., and Spits, H. (2006). Development of human lymphoid cells. Annu Rev Immunol *24*, 287-320.

Buckley, R. H. (2004). Molecular defects in human severe combined immunodeficiency and approaches to immune reconstitution. Annu Rev Immunol *22*, 625-655.

Carson, W. E., Haldar, S., Baiocchi, R. A., Croce, C. M., and Caligiuri, M. A. (1994). The c-kit ligand suppresses apoptosis of human natural killer cells through the upregulation of bcl-2. Proc Natl Acad Sci USA *91*, 7553-7557.

Cooper, M. A., Fehniger, T. A., and Caligiuri, M. A. (2001). The biology of human natural killer-cell subsets. Trends Immunol *22*, 633-640.

De Smedt, M., Hoebeke, I., Reynvoet, K., Leclercq, G., and Plum, J. (2005). Different thresholds of Notch signaling bias human precursor cells toward B-, NK-, monocytic/dendritic-, or T-cell lineage in thymus microenvironment. Blood *106*, 3498-3506.

Diefenbach, A., Jensen, E. R., Jamieson, A. M., and Raulet, D. H. (2001). Rae1 and H60 ligands of the NKG2D receptor stimulate tumour immunity. Nature *413*, 165-171.

Di Santo, J. P. (2006). Natural killer cell developmental pathways: a question of balance. Annu Rev Immunol *24*, 257-286.

Di Santo, J. P., and Vosshenrich, C. A. (2006). Bone marrow versus thymic pathways of natural killer cell development. Immunol Rev *214*, 35-46.

DiSanto, J. P., Muller, W., Guy-Grand, D., Fischer, A., and Rajewsky, K. (1995). Lymphoid development in mice with a targeted deletion of the interleukin 2 receptor gamma chain. Proc Natl Acad Sci USA *92*, 377-381.

Farag, S. S., and Caligiuri, M. A. (2006). Human natural killer cell development and biology. Blood Rev *20*, 123-137.

Fernandez, N. C., Treiner, E., Vance, R. E., Jamieson, A. M., Lemieux, S., and Raulet, D. H. (2005). A subset of natural killer cells achieves self-tolerance without expressing inhibitory receptors specific for self-MHC molecules. Blood *105*, 4416-4423.

Freud, A. G., Becknell, B., Roychowdhury, S., Mao, H. C., Ferketich, A. K., Nuovo, G. J., Hughes, T. L., Marburger, T. B., Sung, J., Baiocchi, R. A. et al. (2005). A human CD34[+] subset resides in lymph nodes and differentiates into CD56bright natural killer cells. Immunity *22*, 295-304.

Freud, A. G., and Caligiuri, M. A. (2006). Human natural killer cell development. Immunol Rev *214*, 56-72.

Furukawa, H., Yabe, T., Watanabe, K., Miyamoto, R., Miki, A., Akaza, T., Tadokoro, K., Tohma, S., Inoue, T., Yamamoto, K., and Juji, T. (1999). Tolerance of NK and LAK activity for HLA class I-deficient targets in a TAP1-deficient patient (bare lymphocyte syndrome type I). Hum Immunol *60*, 32-40.

Galy, A., Travis, M., Cen, D., and Chen, B. (1995). Human T, B, natural killer, and dendritic cells arise from a common bone marrow progenitor cell subset. Immunity *3*, 459-473.

Gimeno, R., Weijer, K., Voordouw, A., Uittenbogaart, C. H., Legrand, N., Alves, N. L., Wijnands, E., Blom, B., and Spits, H. (2004). Monitoring the effect of gene silencing by RNA interference in human CD34+ cells injected into newborn RAG2−/− γ_c−/− mice: functional inactivation of p53 in developing T cells. Blood *104*, 3886-3893.

Gorantla, S., Sneller, H., Walters, L., Sharp, J. G., Pirruccello, S. J., West, J. T., Wood, C., Dewhurst, S., Gendelman, H. E., and Poluektova, L. (2007). HIV-1 pathobiology studied in humanized Balb/c-Rag2−/−γ_c−/− mice. J Virol *81*, 2700-2712.

Haddad, R., Guardiola, P., Izac, B., Thibault, C., Radich, J., Delezoide, A. L., Baillou, C., Lemoine, F. M., Gluckman, J. C., Pflumio, F., and Canque, B. (2004). Molecular characterization of early human T/NK and B-lymphoid progenitor cells in umbilical cord blood. Blood *104*, 3918-3926.

Hanna, J., Bechtel, P., Zhai, Y., Youssef, F., McLachlan, K., and Mandelboim, O. (2004). Novel insights on human NK cells' immunological modalities revealed by gene expression profiling. J Immunol *173*, 6547-6563.

Heemskerk, M. H., Blom, B., Nolan, G., Stegmann, A. P., Bakker, A. Q., Weijer, K., Res, P. C., and Spits, H. (1997). Inhibition of T cell and promotion of natural killer cell development by the dominant negative helix loop helix factor Id3. J Exp Med *186*, 1597-1602.

Ikawa, T., Fujimoto, S., Kawamoto, H., Katsura, Y., and Yokota, Y. (2001). Commitment to natural killer cells requires the helix-loop-helix inhibitor Id2. Proc Natl Acad Sci USA *98*, 5164-5169.

Jaleco, A. C., Blom, B., Res, P., Weijer, K., Lanier, L. L., Phillips, J. H., and Spits, H. (1997). Fetal liver contains committed NK progenitors, but is not a site for development of CD34+ cells into T cells. J Immunol *159*, 694-702.

Karre, K., Ljunggren, H. G., Piontek, G., and Kiessling, R. (1986). Selective rejection of H-2-deficient lymphoma variants suggests alternative immune defence strategy. Nature *319*, 675-678.

Kennedy, M. K., Glaccum, M., Brown, S. N., Butz, E. A., Viney, J. L., Embers, M., Matsuki, N., Charrier, K., Sedger, L., Willis, C. R. et al. (2000). Reversible defects in natural killer and memory CD8 T cell lineages in interleukin 15-deficient mice. J Exp Med *191*, 771-780.

Kim, S., Iizuka, K., Kang, H. S., Dokun, A., French, A. R., Greco, S., and Yokoyama, W. M. (2002). In vivo developmental stages in murine natural killer cell maturation. Nat Immunol *3*, 523-528.

Kim, S., Poursine-Laurent, J., Truscott, S. M., Lybarger, L., Song, Y. J., Yang, L., French, A. R., Sunwoo, J. B., Lemieux, S., Hansen, T. H., and Yokoyama, W. M. (2005). Licensing of natural killer cells by host major histocompatibility complex class I molecules. Nature *436*, 709-713.

Koka, R., Burkett, P. R., Chien, M., Chai, S., Chan, F., Lodolce, J. P., Boone, D. L., and Ma, A. (2003). Interleukin (IL)-15Rα-deficient natural killer cells survive in normal but not IL-15Rα-deficient mice. J Exp Med *197*, 977-984.

Lanier, L. L., Le, A. M., Civin, C. I., Loken, M. R., and Phillips, J. H. (1986). The relationship of CD16 (Leu-11) and Leu-19 (NKH-1) antigen expression on human peripheral blood NK cells and cytotoxic T lymphocytes. J Immunol *136*, 4480-4486.

Lanier, L. L., Testi, R., Bindl, J., and Phillips, J. H. (1989). Identity of Leu-19 (CD56) leukocyte differentiation antigen and neural cell adhesion molecule. J Exp Med *169*, 2233-2238.

Le, P. T., Adams, K. L., Zaya, N., Mathews, H. L., Storkus, W. J., and Ellis, T. M. (2001). Human thymic epithelial cells inhibit IL-15- and IL-2-driven differentiation of NK cells from the early human thymic progenitors. J Immunol *166*, 2194-2201.

Legrand, N., Cupedo, T., van Lent, A. U., Ebeli, M. J., Weijer, K., Hanke, T., and Spits, H. (2006a). Transient accumulation of human mature thymocytes and regulatory T cells with CD28 superagonist in "human immune system" Rag2$^{-/-}\gamma_c^{-/-}$ mice. Blood *108*, 238-245.

Legrand, N., Weijer, K., and Spits, H. (2006b). Experimental models to study development and function of the human immune system in vivo. J Immunol *176*, 2053-2058.

Lodoen, M. B., and Lanier, L. L. (2005). Viral modulation of NK cell immunity. Nat Rev Microbiol *3*, 59-69.

Lodolce, J. P., Boone, D. L., Chai, S., Swain, R. E., Dassopoulos, T., Trettin, S., and Ma, A. (1998). IL-15 receptor maintains lymphoid homeostasis by supporting lymphocyte homing and proliferation. Immunity 9, 669-676.

Lohoff, M., Duncan, G. S., Ferrick, D., Mittrucker, H. W., Bischof, S., Prechtl, S., Rollinghoff, M., Schmitt, E., Pahl, A., and Mak, T. W. (2000). Deficiency in the transcription factor interferon regulatory factor (IRF)-2 leads to severely compromised development of natural killer and T helper type 1 cells. J Exp Med 192, 325-336.

Loza, M. J., and Perussia, B. (2004). The IL-12 signature: NK cell terminal CD56+high stage and effector functions. J Immunol 172, 88-96.

Macchiarini, F., Manz, M. G., Palucka, A. K., and Shultz, L. D. (2005). Humanized mice: are we there yet? J Exp Med 202, 1307-1311.

Martin-Fontecha, A., Thomsen, L. L., Brett, S., Gerard, C., Lipp, M., Lanzavecchia, A., and Sallusto, F. (2004). Induced recruitment of NK cells to lymph nodes provides IFN-γ for $T_H 1$ priming. Nat Immunol 5, 1260-1265.

McCune, J. M. (1997). Animal models of HIV-1 disease. Science 278, 2141-2142.

McCune, J. M., Namikawa, R., Kaneshima, H., Shultz, L. D., Lieberman, M., and Weissman, I. L. (1988). The SCID-hu mouse: murine model for the analysis of human hematolymphoid differentiation and function. Science 241, 1632-1639.

Miller, J. S., Alley, K. A., and McGlave, P. (1994). Differentiation of natural killer (NK) cells from human primitive marrow progenitors in a stroma-based long-term culture system: identification of a CD34+7+ NK progenitor. Blood 83, 2594-2601.

Miller, J. S., and McCullar, V. (2001). Human natural killer cells with polyclonal lectin and immunoglobulinlike receptors develop from single hematopoietic stem cells with preferential expression of NKG2A and KIR2DL2/L3/S2. Blood 98, 705-713.

Morandi, B., Bougras, G., Muller, W. A., Ferlazzo, G., and Munz, C. (2006). NK cells of human secondary lymphoid tissues enhance T cell polarization via IFN-γ secretion. Eur J Immunol 36, 2394-2400.

Mrozek, E., Anderson, P., and Caligiuri, M. A. (1996). Role of interleukin-15 in the development of human CD56+ natural killer cells from CD34+ hematopoietic progenitor cells. Blood 87, 2632-2640.

Namikawa, R., Kaneshima, H., Lieberman, M., Weissman, I. L., and McCune, J. M. (1988). Infection of the SCID-hu mouse by HIV-1. Science 242, 1684-1686.

Nosaka, T., van Deursen, J. M., Tripp, R. A., Thierfelder, W. E., Witthuhn, B. A., McMickle, A. P., Doherty, P. C., Grosveld, G. C., and Ihle, J. N. (1995). Defective lymphoid development in mice lacking Jak3. Science 270, 800-802.

Ogasawara, K., Hida, S., Azimi, N., Tagaya, Y., Sato, T., Yokochi-Fukuda, T., Waldmann, T. A., Taniguchi, T., and Taki, S. (1998). Requirement for IRF-1 in the microenvironment supporting development of natural killer cells. Nature 391, 700-703.

Parham, P. (2006). Taking license with natural killer cell maturation and repertoire development. Immunol Rev 214, 155-160.

Park, S. Y., Saijo, K., Takahashi, T., Osawa, M., Arase, H., Hirayama, N., Miyake, K., Nakauchi, H., Shirasawa, T., and Saito, T. (1995). Developmental defects of lymphoid cells in Jak3 kinase-deficient mice. Immunity 3, 771-782.

Res, P., Martinez-Caceres, E., Cristina Jaleco, A., Staal, F., Noteboom, E., Weijer, K., and Spits, H. (1996). CD34+CD38dim cells in the human thymus can differentiate into T, natural killer, and dendritic cells but are distinct from pluripotent stem cells. Blood 87, 5196-5206.

Roder, J. C., Arlund-Richter, L., and Jondal, M. (1979). Target-effector interaction in the human and murine natural killer system: specificity and xenogeneic reactivity of the solubilized natural killer-target structure complex and its loss in a somatic cell hybrid. J Exp Med 150, 471-481.

Roifman, C. M., Zhang, J., Chitayat, D., and Sharfe, N. (2000). A partial deficiency of interleukin-7Rα is sufficient to abrogate T-cell development and cause severe combined immunodeficiency. Blood 96, 2803-2807.

Rossi, M. I., Yokota, T., Medina, K. L., Garrett, K. P., Comp, P. C., Schipul, A. H., Jr., and Kincade, P. W. (2003). B lymphopoiesis is active throughout human life, but there are developmental age-related changes. Blood *101*, 576-584.
Ryan, D. H., Nuccie, B. L., Ritterman, I., Liesveld, J. L., Abboud, C. N., and Insel, R. A. (1997). Expression of interleukin-7 receptor by lineage-negative human bone marrow progenitors with enhanced lymphoid proliferative potential and B-lineage differentiation capacity. Blood *89*, 929-940.
Samson, S. I., Richard, O., Tavian, M., Ranson, T., Vosshenrich, C. A., Colucci, F., Buer, J., Grosveld, F., Godin, I., and Di Santo, J. P. (2003). GATA-3 promotes maturation, IFN-γ production, and liver-specific homing of NK cells. Immunity *19*, 701-711.
Sarun, S., Dalloul, A. H., Laurent, C., Blanc, C., and Schmitt, C. (1998). Human CD34$^+$ thymocyte maturation: pre-T and NK cell differentiation on neonatal thymic stromal cell culture. Cell Immunol *190*, 23-32.
Sato, T., Laver, J. H., Aiba, Y., and Ogawa, M. (1999). NK cell colony formation from human fetal thymocytes. Exp Hematol *27*, 726-733.
Schotte, R., Nagasawa, M., Weijer, K., Spits, H., and Blom, B. (2004). The ETS transcription factor Spi-B is required for human plasmacytoid dendritic cell development. J Exp Med *200*, 1503-1509.
Schotte, R., Rissoan, M. C., Bendriss-Vermare, N., Bridon, J. M., Duhen, T., Weijer, K., Briere, F., and Spits, H. (2003). The transcription factor Spi-B is expressed in plasmacytoid DC precursors and inhibits T-, B-, and NK-cell development. Blood *101*, 1015-1023.
Spits, H., Blom, B., Jaleco, A. C., Weijer, K., Verschuren, M. C., van Dongen, J. J., Heemskerk, M. H., and Res, P. C. (1998). Early stages in the development of human T, natural killer and thymic dendritic cells. Immunol Rev *165*, 75-86.
Spits, H., Lanier, L. L., and Phillips, J. H. (1995). Development of human T and natural killer cells. Blood *85*, 2654-2670.
Suzuki, H., Duncan, G. S., Takimoto, H., and Mak, T. W. (1997). Abnormal development of intestinal intraepithelial lymphocytes and peripheral natural killer cells in mice lacking the IL-2 receptor beta chain. J Exp Med *185*, 499-505.
Townsend, M. J., Weinmann, A. S., Matsuda, J. L., Salomon, R., Farnham, P. J., Biron, C. A., Gapin, L., and Glimcher, L. H. (2004). T-bet regulates the terminal maturation and homeostasis of NK and Valpha14i NKT cells. Immunity *20*, 477-494.
Traggiai, E., Chicha, L., Mazzucchelli, L., Bronz, L., Piffaretti, J. C., Lanzavecchia, A., and Manz, M. G. (2004). Development of a human adaptive immune system in cord blood cell-transplanted mice. Science *304*, 104-107.
Trinchieri, G. (1989). Biology of natural killer cells. Adv Immunol *47*, 187-376.
Vosshenrich, C. A., Garcia-Ojeda, M. E., Samson-Villeger, S. I., Pasqualetto, V., Enault, L., Richard-Le Goff, O., Corcuff, E., Guy-Grand, D., Rocha, B., Cumano, A. et al. (2006). A thymic pathway of mouse natural killer cell development characterized by expression of GATA-3 and CD127. Nat Immunol *7*, 1217-1224.
Yokoyama, W. M., and Kim, S. (2006). Licensing of natural killer cells by self-major histocompatibility complex class I. Immunol Rev *214*, 143-154.
Yokoyama, W. M., Kim, S., and French, A. R. (2004). The dynamic life of natural killer cells. Annu Rev Immunol *22*, 405-429.
Zhang, L., Kovalev, G. I., and Su, L. (2007). HIV-1 infection and pathogenesis in a novel humanized mouse model. Blood *109*, 2978-2981.

Human T Cell Development and HIV Infection in Human Hemato-Lymphoid System Mice

S. Baenziger, P. Ziegler, L. Mazzucchelli, L. Bronz,
R. F. Speck(✉), and M. G. Manz(✉)

1 Introduction 125
2 Human T Cell Development and HIV Infection in Immunodeficient
   Mice Transplanted as Newborns with Human CD34⁺
   Hematopoietic Stem and Progenitor Cells 126
3 Conclusions 129
References 130

Abstract Advances in generation of mice that on human hematopoietic stem and progenitor cell transplantation develop and maintain human hemato-lymphoid cells have fueled an already thriving field of research. We focus here on human T cell development and HIV infection in $Rag2^{-/-}\gamma_c^{-/-}$ mice transplanted as newborns with human CD34⁺ cord blood hematopoietic stem and progenitor cells.

Abbreviations CCR5: cc-chemokine receptor 5; CXCR: CXC-chemokine receptor; DC: dendritic cell; EBV: Epstein-Barr virus; HIV-1: human immunodeficiency virus type 1; IFN-γ: interferon-gamma; MHC: major histocompatibility complex; MLR: mixed lymphocyte reaction

1 Introduction

For good reasons knowledge of human physiology and pathology is largely gained by observation, cautious, informed consent, and safety-oriented *in vivo* experimentation and *in vitro* surrogate assays. By these approaches, rigorous scientific proof

---

R. F. Speck
Division of Infectious Diseases and Hospital Epidemiology, University Hospital Zurich,
Raemistrasse 100, 8091 Zurich, Switzerland
roberto.speck@usz.ch

M. G. Manz
Institute for Research in Biomedicine (IRB), Via Vincenzo Vela 6, 6500
Bellinzona, Switzerland
manz@irb.unisi.ch

is often impossible. Thus, progress in clinical research is mostly slow. Most societies agree on animal research using worms, flies, and small vertebrates given appropriate ethical consideration. Because of their high similarity with human beings, easy access, and experimental feasibility, laboratory mice have become the main model for *in vivo* biomedical research reported in the vast majority of published work in high-ranking scientific journals. And for sure, the capturing slogan "They (laboratory animals) save more lives than 911 (the emergency call number)" is not an overstatement. However, although genome similarities are higher than naively expected, 65 million years of divergence in human and mouse evolution have shaped two species that differ substantially in size, life span, reproductive activity, and exposure to environmental challenges, as, for example, coevolving species-specific infectious agents. Thus, also regarding hematology and immunology, mice are not men, accounting for one of the reasons that great achievements in mice are quite often lost in translational research [14, 20].

Experimentation with human hemato-lymphoid system mice took off almost 20 years ago, paralleling increasing application of clinical hematopoietic stem cell transplantation for malignant disease and immunodeficiencies, as well as the rising HIV pandemic. Some basic requirements must be met for hematopoietic cell engraftment, differentiation, and function: Cells must find appropriate locations, must be supported by respective nurturing factors from the host environment, and must not be rejected. In cross-species transplantation, this requires immune deficiency of the recipient, as well as cross-reactivity of homing molecules, and differentiation and survival factors, if not produced in sufficient amounts by transplanted cells themselves.

As the detailed history and state of the art in this field were recently reviewed [8, 10, 11, 18] and major scientific contributions to the field are discussed in other chapters of this volume, we here briefly set the main focus of our work on T cell development and HIV infection using immunodeficient $Rag2^{-/-}\gamma_c^{-/-}$ mice (generated by M. Ito at the Central Institute for Experimental Animals, Japan) transplanted as newborns with human $CD34^+$ hematopoietic stem and progenitor cells.

2   Human T Cell Development and HIV Infection in Immunodeficient Mice Transplanted as Newborns with Human $CD34^+$ Hematopoietic Stem and Progenitor Cells

One of the major improvements achieved by using newborn NOD/SCID$\gamma_c^{-/-}$ and $Rag2^{-/-}\gamma_c^{-/-}$ mice as recipients for human hematopoietic stem and progenitor cell grafts is efficient intrathymic *de novo* development of human T cells [3, 5, 7, 23]. In contrast to newborn transplantation, T cell development in adult NOD/SCID$\gamma_c^{-/-}$ recipients is less efficient but can be enhanced by exogenously adding human IL-7 [6, 19]. Human T cells generated in the mouse thymus include in fairly physiological ratios $CD4^+$ and $CD8^+$ T cells with a broad Vβ distribution,

Foxp3[+] CD25[+] regulatory T cells, and γδ T cells. Mature T cells exit the thymus and home to secondary lymphoid organs. What is the MHC restriction and functionality of human T cells educated on a mouse thymic background? Data generated mostly in mice demonstrate that under normal developmental conditions positive selection preferentially occurs on cortical thymic epithelial cells, while both medullary epithelial cells and hematopoietic derived dendritic cells are involved in negative selection (e.g., reviewed in [15]). We did not observe human thymic epithelial cells in CD34[+] cell-transplanted mice in line with expectations of epithelial cell germ layer derivation, while dendritic cells of both mouse and human origin constituted the thymus. Thus, human developing thymocytes should in theory be positively selected on mouse epithelial cells, while negative selection might occur on both mouse and human MHC. If so, it would be reasonable to expect that T cells positively selected on mouse MHC would continue to preferentially interact in negative selection also with mouse MHC. However, under certain conditions hematopoietic offspring cells are likely involved in positive selection, as pointed out by several mouse-to-mouse transplantation studies [17, 26]. Moreover, species-specific differences in thymic selection might exist, and human thymocyte-thymocyte MHC class II interaction, and thus at least some CD4[+] T cell selection, might occur on human MHC [4, 9]. On T cell exit from the thymus, they depend on homeostatic factors, both MHC and cytokines for survival (e.g., reviewed in [22]). As in human hematopoietic stem and progenitor cell-transplanted mice human MHC is only present on hematopoietic but no other tissues, presentation of both "self" and "nonself" peptides in the context of both MHC class I and II will depend on the cellular tropism of these.

Given all the considerations mentioned above, what experimental data on reactivity of mouse background-generated human T cells has been reported? (1) Human T cells isolated from mouse lymph nodes and spleen proliferated vigorously in mixed lymphocyte reactions (MLRs) when stimulated with human allogeneic DCs, but weakly or not at all when stimulated with human autologous DCs. Proliferative response to mouse DCs was low; however, there was a small difference with stronger proliferation of T cells when stimulated with fully mismatched versus host mouse type DC [23]. (2) Cytotoxic activity against human allogeneic target cells could be blocked with human MHC class I or II antibodies, respectively [7]. (3) Some responses of human T cells in mice to *in vivo* infection of human B cells with Epstein-Barr virus (EBV) are observed; however, in some cases human T cells were not capable to control EBV-driven B cell proliferation [23]. (4) Human T cells specific for viral epitopes were only observed in the context of mouse MHC on infection of mice with influenza virus [8]. (5) When mature mouse-derived human T cells were transferred to nontransplanted Rag2$^{-/-}$γ$_c^{-/-}$ mice, we observed no relevant homeostatic cell expansion; similarly, high peripheral T cell turnover rates and lack of long-term T cell maintenance were observed by other groups [8], suggesting that peripheral homeostatic T cell maintenance might not function appropriately, an issue that still needs to be resolved by more experimentation.

In summary, thus far no firm conclusions can be drawn on the biology of T cell selection in this setting where the T cells are of human origin and the thymic stroma is of mouse origin. It can only be suggested that T cell tolerance, that is, possibly negative selection, for both autologous human and mouse MHC is achieved. Beyond that, insufficient T cell responses due to no or only weakly cross-species reactive costimulation and cytokine responses might account for additional difficulties in any human T cell response observed in this setting. In terms of appropriate T cell selection, the obvious solution is to replace mouse by human MHC components, creating at least for thymic selection a similar situation as in mice cotransplanted with same donor fetal thymic tissue [12, 13]. It will be of great interest to see whether human T cell development will progress in the absence of any tissue MHC, a model situation of allogeneic hematopoietic cell transplantation in human MHC deficiencies [16].

As in our human cord blood $CD34^+$ cell-transplanted mice human T cells developed and seeded secondary lymphoid organs, we, and concomitantly others, tested these mice and mice cotransplanted with human thymus as models for HIV infection [1, 2, 21, 24, 25]. We infected human hemato-lymphoid system mice intraperitoneally with either CCR5-tropic or CXCR4-tropic HIV-1 strains. Irrespective of coreceptor selectivity, HIV RNA plasma copies peaked at 2-4 weeks after infection, comparable to HIV infection in humans. Thereafter, viremia mostly stabilized at lower levels and was maintained for up to 190 days, the longest time followed. HIV generated in mice was functional, since supernatants of infected mouse-derived cell cultures propagated infection in primary human leukocytes. As in humans, developing human $CD4^+$ thymocytes in mice are mostly CXCR4-positive, but lack CCR5 expression, while peripheral $CD4^+$ T cells express CXCR4 and/or CCR5. Thymic HIV infection was detected on CXCR4-tropic infection, while secondary lymphoid organ infection occurred in both CXCR4- and CCR5-tropic virus-infected animals (Fig. 1). In both CXCR4- and CCR5-tropic strain-infected animals, productively HIV-infected cells, namely, HIV-$p24^+$ cells, were mostly $CD3^+$ and only occasionally non-T cells such as $CD68^+$ macrophages. In some mice with high numbers of productively infected cells, syncytium formation occurred in both spleen and lymph nodes, a process observed in brains and lymphoid tissue of HIV-infected individuals, likely associated with high viral replication, spreading, and $CD4^+$ cell loss. Overall, CXCR4-tropic HIV infection led to more rapid blood $CD4^+$ cell loss than CCR5-tropic infection, reminiscent of CXCR4-tropic emergence of HIV strains in late-stage human HIV disease. One of 25 infected mice mounted a HIV-specific IgG response detectable by standard clinical assays [1]. Furthermore, although based on limited data, we did not detect robust HIV-specific T cell responses, as determined by IFN-γ detection on *in vitro* restimulation, in line with other concomitant reports [24]. Thus, although specific immune responses are observed, they thus far lack robustness, that is, they are not predictable in frequency and levels, prohibiting at this stage, for example, efficient preclinical vaccine testing. However, as it stands, this model will be valuable to study virus-induced pathology and to evaluate new nonadaptive immunity-dependent approaches aiming to fight HIV.

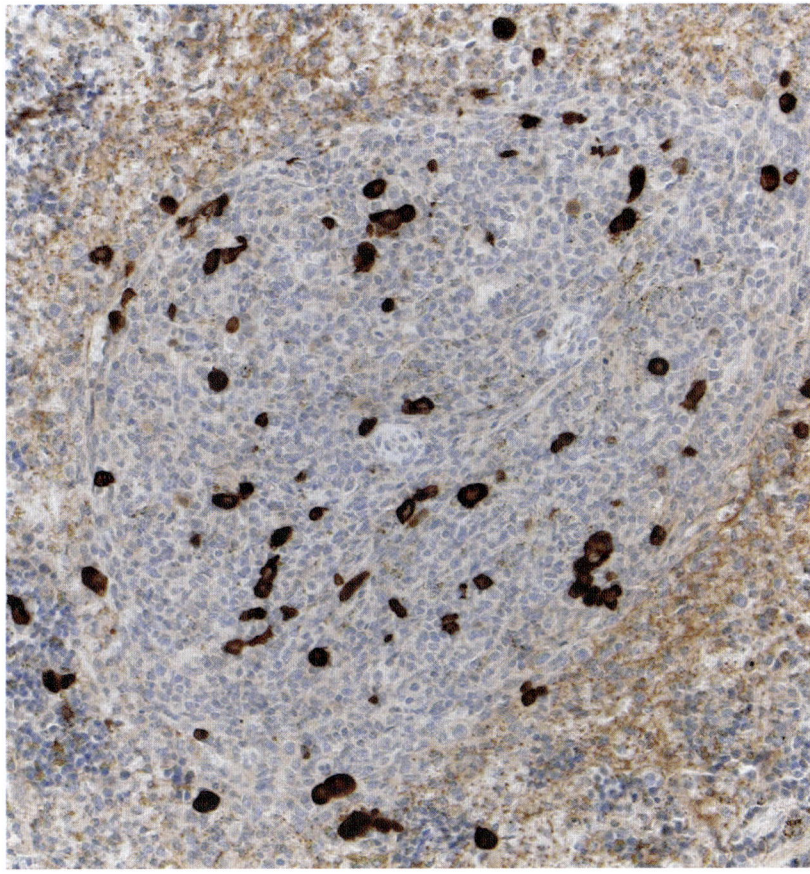

Fig. 1 Rag2$^{-/-}\gamma_c^{-/-}$ mice, transplanted as newborns with human CD34$^+$ cord blood cells and subsequently infected with HIV, develop lymphoid organ-disseminated, productive HIV infection. Histology shows HIV p24 staining preferentially localized in the white pulp area in the spleen of an HIV-infected mouse (18 days after infection with CCR5-tropic YU-2; Magnification 20x)

3 Conclusions

It can be anticipated that basic and preclinical *in vivo* human immunology and infectious disease research will greatly benefit from improved, easy-to-generate, and broadly available human hemato-lymphoid system mice over the coming years.

Acknowledgements We would like to thank the staff of the department of Obstetrics, Ospedale San Giovanni for cord blood supply, M. Ito for provision of BALB/c Rag2$^{-/-}\gamma_c^{-/-}$ mice, and the Swiss National Science Foundation and the Bill and Melinda Gates Foundation for research support.

References

1. Baenziger, S., R. Tussiwand, E. Schlaepfer, L. Mazzucchelli, M. Heikenwalder, M.O. Kurrer, S. Behnke, J. Frey, A. Oxenius, H. Joller, A. Aguzzi, M.G. Manz, and R.F. Speck. 2006. Disseminated and sustained HIV infection in CD34+ cord blood cell-transplanted Rag2−/−gamma c−/− mice. *Proc Natl Acad Sci USA* 103:15951-15956.
2. Berges, B.K., W.H. Wheat, B.E. Palmer, E. Connick, and R. Akkina. 2006. HIV-1 infection and CD4 T cell depletion in the humanized Rag2−/−gamma c−/− (RAG-hu) mouse model. *Retrovirology* 3:76.
3. Chicha, L., R. Tussiwand, E. Traggiai, L. Mazzucchelli, L. Bronz, J.C. Piffaretti, A. Lanzavecchia, and M.G. Manz. 2005. Human adaptive immune system Rag2−/−γc−/− Mice. *Ann NY Acad Sci* 1044:236-243.
4. Choi, E.Y., K.C. Jung, H.J. Park, D.H. Chung, J.S. Song, S.D. Yang, E. Simpson, and S.H. Park. 2005. Thymocyte-thymocyte interaction for efficient positive selection and maturation of CD4 T cells. *Immunity* 23:387-396.
5. Gimeno, R., K. Weijer, A. Voordouw, C.H. Uittenbogaart, N. Legrand, N.L. Alves, E. Wijnands, B. Blom, and H. Spits. 2004. Monitoring the effect of gene silencing by RNA interference in human CD34+ cells injected into newborn RAG2−/− gammac−/− mice: functional inactivation of p53 in developing T cells. *Blood* 104:3886-3893.
6. Hiramatsu, H., R. Nishikomori, T. Heike, M. Ito, K. Kobayashi, K. Katamura, and T. Nakahata. 2003. Complete reconstitution of human lymphocytes from cord blood CD34+ cells using the NOD/SCID/gammacnull mice model. *Blood* 102:873-880.
7. Ishikawa, F., M. Yasukawa, B. Lyons, S. Yoshida, T. Miyamoto, G. Yoshimoto, T. Watanabe, K. Akashi, L.D. Shultz, and M. Harada. 2005. Development of functional human blood and immune systems in NOD/SCID/IL2 receptor γ chain^{null} mice. *Blood* 106:1565-1573.
8. Legrand, N., K. Weijer, and H. Spits. 2006. Experimental models to study development and function of the human immune system in vivo. *J Immunol* 176:2053-2058.
9. Li, W., M.G. Kim, T.S. Gourley, B.P. McCarthy, D.B. Sant'Angelo, and C.H. Chang. 2005. An alternate pathway for CD4 T cell development: thymocyte-expressed MHC class II selects a distinct T cell population. *Immunity* 23:375-386.
10. Macchiarini, F., M.G. Manz, A.K. Palucka, and L.D. Shultz. 2005. Humanized mice: are we there yet? *J Exp Med* 202:1307-1311.
11. Manz, M.G. 2007. Human-hemato-lymphoid-system mice: opportunities and challenges. *Immunity* 26:537-541.
12. McCune, J.M., R. Namikawa, H. Kaneshima, L.D. Shultz, M. Lieberman, and I.L. Weissman. 1988. The SCID-hu mouse: murine model for the analysis of human hematolymphoid differentiation and function. *Science* 241:1632-1639.
13. Melkus, M.W., J.D. Estes, A. Padgett-Thomas, J. Gatlin, P.W. Denton, F.A. Othieno, A.K. Wege, A.T. Haase, and J.V. Garcia. 2006. Humanized mice mount specific adaptive and innate immune responses to EBV and TSST-1. *Nat Med* 12:1316-1322.
14. Mestas, J., and C.C. Hughes. 2004. Of mice and not men: differences between mouse and human immunology. *J Immunol* 172:2731-2738.
15. Palmer, E. 2003. Negative selection-clearing out the bad apples from the T-cell repertoire. *Nat Rev Immunol* 3:383-391.
16. Reith, W., and B. Mach. 2001. The bare lymphocyte syndrome and the regulation of MHC expression. *Annu Rev Immunol* 19:331-373.
17. Shizuru, J.A., I.L. Weissman, R. Kernoff, M. Masek, and Y.C. Scheffold. 2000. Purified hematopoietic stem cell grafts induce tolerance to alloantigens and can mediate positive and negative T cell selection. *Proc Natl Acad Sci USA* 97:9555-9560.
18. Shultz, L.D., F. Ishikawa, and D.L. Greiner. 2007. Humanized mice in translational biomedical research. *Nat Rev Immunol* 7:118-130.
19. Shultz, L.D., B.L. Lyons, L.M. Burzenski, B. Gott, X. Chen, S. Chaleff, M. Kotb, S.D. Gillies, M. King, J. Mangada, D.L. Greiner, and R. Handgretinger. 2005. Human lymphoid and

myeloid cell development in NOD/LtSz-scid IL2R gamma null mice engrafted with mobilized human hemopoietic stem cells. *J Immunol* 174:6477-6489.
20. Steinman, R.M., and I. Mellman. 2004. Immunotherapy: bewitched, bothered, and bewildered no more. *Science* 305:197-200.
21. Sun, Z., P.W. Denton, J.D. Estes, F.A. Othieno, B.L. Wei, A.K. Wege, M.W. Melkus, A. Padgett-Thomas, M. Zupancic, A.T. Haase, and J.V. Garcia. 2007. Intrarectal transmission, systemic infection, and CD4+ T cell depletion in humanized mice infected with HIV-1. *J Exp Med* 204:705-714.
22. Surh, C.D., and J. Sprent. 2005. Regulation of mature T cell homeostasis. *Semin Immunol* 17:183-191.
23. Traggiai, E., L. Chicha, L. Mazzucchelli, L. Bronz, J.C. Piffaretti, A. Lanzavecchia, and M.G. Manz. 2004. Development of a human adaptive immune system in cord blood cell-transplanted mice. *Science* 304:104-107.
24. Watanabe, S., K. Terashima, S. Ohta, S. Horibata, M. Yajima, Y. Shiozawa, M.Z. Dewan, Z. Yu, M. Ito, T. Morio, N. Shimizu, M. Honda, and N. Yamamoto. 2007. Hematopoietic stem cell-engrafted NOD/SCID/IL2Rgamma null mice develop human lymphoid systems and induce long-lasting HIV-1 infection with specific humoral immune responses. *Blood* 109:212-218.
25. Zhang, L., G.I. Kovalev, and L. Su. 2007. HIV-1 infection and pathogenesis in a novel humanized mouse model. *Blood* 109:2978-2981.
26. Zinkernagel, R.M., and A. Althage. 1999. On the role of thymic epithelium vs. bone marrow-derived cells in repertoire selection of T cells. *Proc Natl Acad Sci USA* 96:8092-8097.

Humanized Mice for Human Retrovirus Infection

Y. Koyanagi(✉), Y. Tanaka, M. Ito, and N. Yamamoto

1	Mouse/Human Chimeric Models for HIV-1 Infection Using SCID Mouse	134
	1.1 HIV-1 Infection in the SCID-hu thy/liv Mouse	134
	1.2 HIV-1 Infection in the hu-PBL-SCID Mouse	136
	1.3 Human Acquired Immune Responses in the hu-PBL-SCID Mouse	136
2	Development of Immunodeficient Mouse Strains for Improvement of Human Cell Reconstitution and HIV-1 Infection	138
	2.1 Generation of Novel Immunodeficient Mouse Strains	138
	2.2 HIV-1 Infection in the hu-PBL-NOG Mouse	139
	2.3 Development of HSC-Engrafted Mouse for HIV-1 Infection	140
3	Application of NOG Mice for HTLV Infection	141
	3.1 HTLV	141
	3.2 Model of HTLV-I-Induced Tumorigenicity	141
	3.3 Evaluation of Anti-ATL Compounds	142
4	Concluding Remarks on Human Retrovirus Infection Model	143
	References	144

Abstract Inbred mice with specific genetic defects have greatly facilitated the analysis of complex biological events. Several humanized mouse models using the C.B.-17 *scid/scid* mouse (referred to as the SCID mouse) have been created from two transplantation protocols, and these mice have been utilized for the investigation of human immunodeficiency virus type 1 (HIV-1) and human T-lymphotropic virus type I (HTLV-I) pathogenesis and the evaluation of antiviral compounds. To generate a more prominent small animal model for human retrovirus infection, especially for examination of the pathological process and the immune reaction, a novel immunodeficient mouse strain derived from the NOD SCID mouse was created by backcrossing with a common γ chain (γ_c)-knockout mouse. The NOD-SCID γ_c^{null} (NOG) mouse has neither functional T and B cells nor NK cells and has been used as a recipient in humanized mouse models

Y. Koyanagi
Laboratory of Viral Pathogenesis, Institute for Virus Research, Kyoto University,
53 Shougoinkawara cho, Sakyou-ku, Kyoto 606-8507, Japan
ykoyanag@virus.kyoto-u.ac.jp

for transplantation of human immune cells particularly including hematopoietic stem cells (HSC). From recent advances in development of humanized mice, we are now able to provide a new version of the animal model for human retrovirus infection and human immunity.

Abbreviations AIDS: acquired immunodeficiency syndrome; APC: antigen-presenting cell; ATL: adult T-cell leukemia; AZT: azidothymidine; CCR5: cc-chemokine receptor 5; DC: dendritic cell; ddI: dideoxyinosine; DN: double negative; γ_c; common gamma chain; GVHD: graft-versus-host disease; HAM: HTLV-I-associated myelopathy; HIV-1: human immunodeficiency virus type 1; HSC: hematopoietic stem cell; HTLV-I: human T-lymphotropic virus type I; IL: interleukin; MHC: major histocompatibility complex; PBMC: peripheral blood mononuclear cell; SCID: severe combined immunodeficiency; SP: single positive; TRAIL: tumor necrosis factor (TNF)-related apoptosis-inducing ligand; TSP: tropical spastic paraparesis

1 Mouse/Human Chimeric Models for HIV-1 Infection Using SCID Mouse

The C.B.-17 *scid/scid* (SCID) mouse carries a spontaneously arising autosomal recessive mutation and was found to have severe combined immunodeficiency (SCID) by Bosma and colleagues (Bosma et al. 1983). This strain has a defect of DNA protein kinase and a lack of progenitors to T and B cells (Blunt et al. 1995; Boubnov and Weaver 1995; Kirchgessner et al. 1995; Miller et al. 1995; Peterson et al. 1995). Therefore, these mice are unable to reject the xenograft and they tolerate engraftment of human cells or tissues, followed by subsequent infection with human immunodeficiency virus type 1 (HIV-1). Two of these models are the SCID-hu thy/liv mouse developed by McCune and colleagues (McCune et al. 1988) and, the hu-PBL-SCID mouse developed by Mosier and colleagues (Mosier et al. 1988) (Fig. 1).

1.1 HIV-1 Infection in the SCID-hu thy/liv Mouse

The SCID-hu thy/liv mouse is generated by surgical coimplantation of a piece (1 mm) of human fetal thymus and liver under the murine kidney capsule. The implanted tissues produce a conjoint organ (thy/liv) that appears to reconstitute normal thymus for more than 1 year. The fetal liver provides hematopoietic precursors that are located in islets between the thymic lobes. Thymic epithelial cells are derived from the fetal thymus, whereas the hematopoietic cells including T cells and dendritic cells (DC) are derived from the liver (Vandekerckhove et al. 1992). The generated thymus is composed of more than 70% CD4CD8

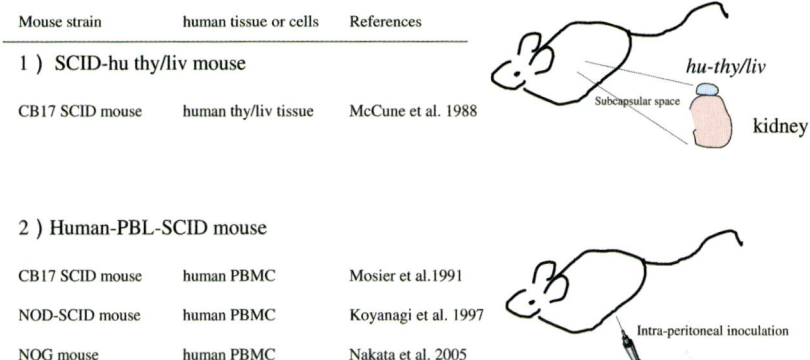

Mouse strain	human tissue or cells	References
1) SCID-hu thy/liv mouse		
CB17 SCID mouse	human thy/liv tissue	McCune et al. 1988
2) Human-PBL-SCID mouse		
CB17 SCID mouse	human PBMC	Mosier et al.1991
NOD-SCID mouse	human PBMC	Koyanagi et al. 1997
NOG mouse	human PBMC	Nakata et al. 2005

Fig. 1 Mouse/human chimeric models for HIV-1 infection. HIV-1-susceptible humanized mouse models have been reported from two transplantation protocols

double-positive (DP) cells, and the rest are CD4 or CD8 single-positive (SP) and double-negative (DN) T cells. Although these CD4 or CD8 SP cells migrate into the peripheral circulation and express a naive phenotype (>70% CD45RA$^+$), the numbers of these cells are very low (Jamieson and Zack 1999). In addition, no immune response to viral antigen has been found so far. Thus, the reconstitution of the human immune system is not complete in this mouse. Since some skill in the surgical operation to coimplant the thy/liv organ and systematic support on distribution of human fetal organs are required, the SCID-hu thy/liv mouse studies have been carried out in limited numbers of laboratories, mainly in the US.

The human thy/liv implant in SCID mice is highly susceptible to infection with HIV-1, including R5 and X4 HIV-1 (McCune et al. 1991; Namikawa et al. 1988). As the number of circulating human CD4$^+$ T cells is low as mentioned above, direct injection of virus into the implant is performed under anesthesia. The level of HIV-1 replication and potential with CD4 T cell depletion of X4 HIV-1 appears to be higher than that of R5 HIV-1 within a few weeks of infection by evaluation with PCR and flow cytometric analyses (Kaneshima et al. 1994). Immunohistological examination indicated that the infected cells initially appeared in the thymic cortical regions and subsequently spread through the entire organ after HIV-1 infection (Stanley et al. 1993). Flow cytometric analysis indicated that the CD4CD8 DP cells were almost completely eradicated and the ratio of CD4$^+$ and CD8$^+$ thymocytes was reversed after infection (Aldrovandi et al. 1993; Bonyhadi et al. 1993). Furthermore, SCID-hu thy/liv mouse have been also used to assess efficacy of several anti-HIV compounds administrated before infection, including azidothymidine (AZT) and 2′,3′-dideoxyinosine (ddI) (Kaneshima et al. 1991; McCune et al. 1990; Rabin et al. 1996).

1.2 HIV-1 Infection in the hu-PBL-SCID Mouse

The hu-PBL-SCID mouse is created by injection of peripheral blood mononuclear cells (PBMC) from healthy adults into the peritoneal cavity of SCID mice, and human $CD4^+$ and $CD8^+$ T cells circulate through the peritoneal cavity, peripheral blood, and organs such as spleen and liver for more than 1 month after PBMC injection (Mosier et al. 1988). The presence of human $CD4^+$ T cells makes this attractive as a model to study HIV-1 pathogenesis and evaluation of anti-HIV compounds (Mosier et al. 1991; Pastore et al. 2003; Hartley et al. 2004). Although it was initially reported as a model with little graft-versus-host disease (GVHD) in the hu-PBL-SCID mice, the high levels of T cell-reconstituted mice develop symptoms of GVHD within 2 months after injection of PBMC (Sandhu et al. 1995). Therefore, long-term observation may not be possible in this model. In addition, human $CD4^+$ and $CD8^+$ T cells in hu-PBL-SCID mice expressed the CD45RO antigen, a marker found in either activated or memory T cells (Tary-Lehmann and Saxon 1992). This is not similar to the ratio of normal adult PBMC, which contain approximately 50% $CD45RO^+$ and $CD45RA^+$ (naive) cells. Furthermore, human $CD4^+$ cells in the hu-PBL-SCID also express abundant levels of HIV-1 coreceptor CCR5, and accordingly, R5 but not X4 HIV-1 more actively replicates in this system (Mosier et al. 1993; Nakata et al. 2006). The relative ease with which this model can be generated and the high efficiency of R5 HIV-1 infection make this system very attractive to study HIV-1 pathogenesis for researchers who struggle in obtaining fetal organs for generating SCID-hu thy/liv mouse. Importantly, a significantly high level of HIV-1 replication correlates with severe depletion of human $CD4^+$ T cells within 2 weeks after infection, and the replication is dependent on Nef protein (Kawano et al. 1997). Thus, this model appears to be adequate for short-term investigation of HIV-1 replication and pathogenesis. As mentioned above, the reconstituted $CD4^+$ and $CD8^+$ T cells are strongly activated and have memory phenotypes, indicating that these cells are xenoreactive proliferated but anergic T cells (Tary-Lehmann and Saxon 1992; Tary-Lehmann at al. 1995). In this model, neither thymopoiesis nor hematopoiesis is generated (Koyanagi, unpublished observations). Thus, it is assumed that the pathological events of HIV-1 infection in this model include that in mature T cells in humans.

1.3 Human Acquired Immune Responses in the hu-PBL-SCID Mouse

Of interest, some immune responses are induced, including humoral and cellular immune responses in the hu-PBL-SCID mouse with administration of various antigens (Gorantla et al. 2005; Ifversen P and Borrebaeck 1996; Nonoyama et al. 1993; Sandhu et al. 1994). However, there are two major limitations to development of strong human immune responses in the hu-PBL-SCID mice. The first is the lack of appropriate human antigen-presenting cells (APC) including DC. The second is the lack of a suitable microenvironment such as the presence of normal lymphoid organs and architecture that may facilitate induction and maintenance of immune effector cells (Tary-Lehmann et al. 1995). To overcome the lack of APC, Delhem et al. used autologous skin that contains tissue DC as a source of APC and succeeded in demonstrating the induction of primary MHC-restricted human T cell responses against HIV-1 envelope in the hu-PBL-SCID mice (Delhem et al. 1998). Santini et al. have reported that inactivated HIV-1-pulsed, monocyte-derived, and matured human DC can stimulate human anti-HIV-1 antibody production by B cells from HIV-1-negative donors in the SCID mouse system, and that this immune response is partially protective (Santini et al. 2000). However, there had been no attempts to overcome the lack of a suitable microenvironment in this hu-PBL-SCID mouse until our report (Yoshida et al. 2003).

To overcome the deficiency of a suitable microenvironment in the hu-PBL-SCID mouse system, we attempted to transfer PBMC together with inactivated HIV-1-pulsed autologous monocyte-derived DC directly into the mouse spleen (Yoshida et al. 2003). The intrasplenic inoculation of PBMC was found to reduce excessive GVHD compared to the intraperitoneal transfer method (Tanaka et al., unpublished observations). In addition, with this procedure we could obtain larger yields of human T cells than with the conventional intraperitoneal transfer. Therefore, we reasoned that the microenvironment in the mouse spleen should provide human T cells with optimum conditions for activation (Fig. 2). With this new immunizing protocol, we have succeeded in eliciting a protective $CD4^+$ T cell immunity against R5 HIV-1 infection. We were surprised to see that the immunized mice were totally resistant against challenge with live R5 HIV-1 (Yoshida et al. 2003). The protective immunity was induced equally with either R5 or X4 HIV-1 as an antigen. The sera from the immunized mice contained a soluble R5 HIV-1 suppressive factor that was mainly synthesized by human $CD4^+$ T cells in response to HIV-1 antigen, specific peptides of HIV-1 according to MHC class II haplotypes (Yoshida et al. 2005), but

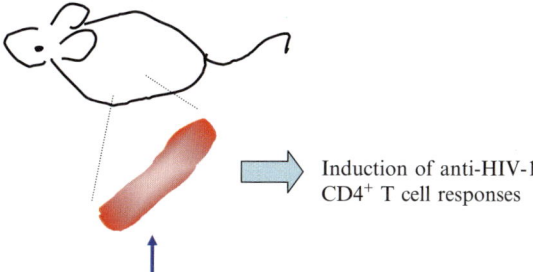

Fig. 2 Human acquired immune responses in the hu-PBL-SCID mouse. Protective $CD4^+$ T cell immunity against R5 HIV-1 infection is induced by PBMC transfer together with inactivated HIV-1-pulsed autologous DC directly into the mouse spleen

not ovalbumin as a control. The factor appears to be unrelated to the currently identified R5 HIV-1 suppressive cytokines examined by neutralization assay using antibodies against CCR5-binding β-chemokines, IL-4, IL-10, IL-12, IL-13, IL-16, MCP-1, MCP-3, IFN-α, IFN-β, TNF-α, and TNF-β.

Our study indicates that a DC-based HIV-1 vaccination can induce HIV-1-reactive human $CD4^+$ T cells producing a yet-undefined R5 HIV-1 suppressor factor. The demonstration made by Lu et al. (Lu et al. 2003) that DC pulsed with AT-2 inactivated simian immunodeficiency virus (SIV) can also stimulate protective anti-SIV specific T cell and antibody responses in rhesus monkeys suggests a rational basis for the DC-based immunization against HIV-1 infection. This idea is further supported by the recent findings by Lu et al. (Lu et al. 2004), who showed the efficacy of a therapeutic DC-based whole HIV-1 virion vaccination for HIV-1 infection.

In order to achieve successful DC-based immunization against HIV-1, a large number of monocyte-derived mature DC with immuno-stimulating activity is expected. A conventional method for generating DC from monocytes has been established, and there are commercially available kits for monocyte purification. In addition, we recently found a simple method to isolate monocytes from bulk PBMC by using antichemokine receptor monoclonal antibody-coated plates (Nimura et al. 2006). These monocytes could be differentiated to Th1-inducing DC and expressed low levels of cell surface CD4 and CCR5. When sensitized with inactivated HIV-1, these DC could induce the R5 HIV-1-suppressing factor in the hu-PBL-SCID mouse (Nimura et al. 2006). Because human monocytes and immature DC are still susceptible to R5 HIV-1 infection, this method, which can induce HIV-1-resistant DC, will be helpful for HIV-1-infected individuals in generation of therapeutic DC against acquired immunodeficiency syndrome (AIDS).

2 Development of Immunodeficient Mouse Strains for Improvement of Human Cell Reconstitution and HIV-1 Infection

2.1 Generation of Novel Immunodeficient Mouse Strains

By introduction of the *scid/scid* mutation in various strains of inbred mice possessing defects in innate immunity, we surveyed congenic mouse strains to find an adequate immunodeficient mouse for transplantation of human cells and HIV-1 infection. From extensive transplantation experiments with human PBMC in immunodeficient mouse strains, the NOD-SCID mouse, which contains some defects in the function of complements and macrophages, was found to be a most suitable strain as a recipient for human cell reconstitution as well as HIV-1 infection (Koyanagi et al. 1997a). The number of human cells in the human PBMC-transplanted NOD-SCID (hu-PBL-NOD-SCID) mice was more than three- to fivefold higher than that in the conventional hu-PBL-SCID mice. More importantly, the levels of HIV-1 viremia were significantly high in the HIV-1 infected hu-PBL-NOD-SCID mice. Some HIV-1-infected mice exhibited more than 1 ng of p24 antigen per milliliter in plasma, which is a much higher concentration than that in HIV-1-infected patients, and showed severe CD4 depletion after intraperitoneal inoculation with R5 HIV-1. Using this mouse model, we found that tumor necrosis factor (TNF)-related apoptosis-inducing ligand (TRAIL) but not Fas ligand induced bystander killing of CD4⁺ T cells in the mouse spleen (Miura et al. 2001). In addition, when these HIV-1-infected mice were treated with a sublethal dose of lipopolysaccharide, HIV-1-infected cells migrated into mouse brain tissue and induced apoptosis in neuronal cells via TRAIL molecules expressed on the HIV-1-infected cells, probably macrophages (Miura et al. 2003). These data suggested that TRAIL should have a pathological role of disease progression for AIDS.

Recently, further genetic manipulation has been performed. A more profoundly severe immunodeficient mouse strain, defective in common γ (γ_c) chain, which is a component of receptors for IL-2, IL-7, and other cytokines and critical for generation of NK and T cells, was generated from the NOD-SCID mouse (Ito et al. 2002) and the recombination activating gene-2 (RAG-2)^{null} mouse. The RAG-2^{null} mouse is also a genetically manipulated immunodeficient strain, which has a defect in the differentiation of T and B cells. NOD-SCID γ_c^{null} (NOG) and RAG-2/γ_c^{null} mouse strains were then generated, and these mice have been confirmed to have neither functional T and B cell nor NK cells and have been used as humanized mouse models by transplantation of human immune cells. The level of human cell reconstitution was greatly improved in NOG as well as RAG-2/γ_c^{null} mice transplanted with human PBMC (hu-PBL-NOG and hu-PBL-RAG-2/γ_c^{null} mice) (Nakata et al. 2005; Koyanagi et al. 1997b). Moreover, a technical improvement should be noted in that it is not necessary for preceding treatment with antibody, an anti-IL-2 receptor β chain antibody or an anti-asialo GM-1 antibody, to protect mouse NK cell differentiation in recipient mice.

2.2 HIV-1 Infection in the hu-PBL-NOG Mouse

Using NOG mice, we indicated the usefulness of the mouse strain for evaluation of an anti-HIV-1 compound (Nakata et al. 2006). HIV-1 suppressive efficacy of a small molecule of CCR5 antagonist was confirmed in the mice 1 day after HIV-1 inoculation. The level of viral loads and HIV-1-induced CD4 depletion was dramatically suppressed after treatment of a CCR5 antagonist, suggesting that the NOG mouse should serve as a small animal model for evaluation of anti-HIV-1 compounds (Nakata at al. 2006). Furthermore, it is noteworthy that this hu-PBL-NOG model provides a greater reproducibility of high viremia levels than the conventional HIV-1-infected SCID mouse models and that the high level of viremia achieved in this mouse model made it possible to monitor the changes in the viremia levels periodically in the same set of mice without sacrificing them, while most of the previously described SCID mouse models required mice to be sacrificed at each time point of testing (Mosier et al. 1993; Ruxrungtham et al. 1996; Strizki et al. 2001).

2.3 Development of HSC-Engrafted Mouse for HIV-1 Infection

Since human immune cells are not fully reconstituted in the chimeric mice transplanted with human PBMC, many researchers have been developing a improved immunodeficient mouse model engrafted with human HSC, which generates human T and B cells and also DC/macrophage myeloid cells for long periods of time and will be amenable to HIV-1 infection (Fig. 3). The HSC-engrafted mouse model allows mechanistic studies of HIV-1-induced disease progression, and possibly would allow analysis of immune responses derived from the human cells.

Using 8- to 10-week-old NOG mice engrafted with HSC or progenitor cells from cord blood, Nakahata and colleagues reported the significantly efficient reconstitution of human B cells as well as T cells in mice for more than 6 months (Hiramatsu et al. 2003). This successful engraftment (designated as the hNOG mouse) encouraged the subsequent HIV-1 infection experiment. We recently showed that in the hNOG mouse exhibiting long-lasting high levels of viremia and CD4 depletion in the peripheral blood were reproduced after both R5 and X4 HIV-1

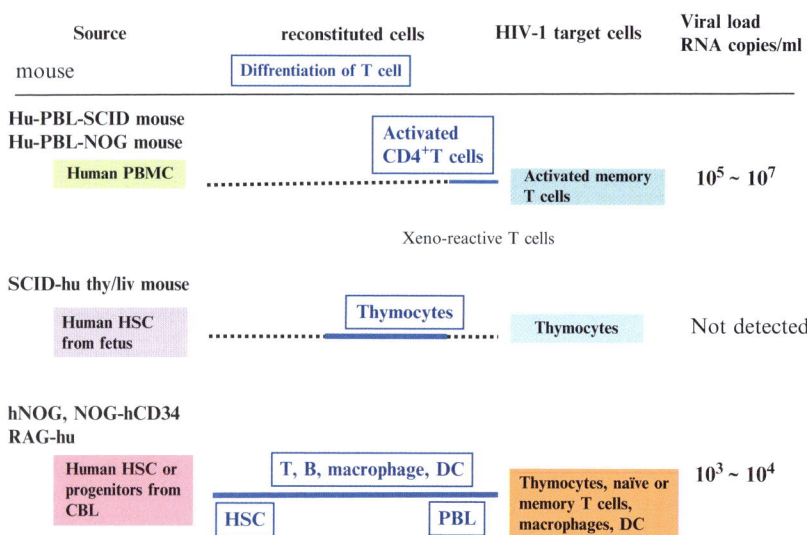

Fig. 3 Comparison of mouse/human chimeric models for HIV-1 infection. There have been three systems of humanized mouse models for HIV-1 infection

inoculation. Large amounts of HIV-1 DNA were detected in the spleen and bone marrow of R5 HIV-1-infected mice, and in the thymus and spleen of X4 virus-infected mice. Furthermore, anti-HIV p24 and gp120 antibodies were found in animals showing a high level of HIV-1 replication, indicating that HIV-1-specific immune response is induced in hNOG mice (Watanabe et al. 2007).

One modification of the CD34 engraftment into mice has been developed. Manz and colleagues reported that the newborn immunodeficient RAG-2/γ_c^{null} mouse strain is an adequate recipient for reconstitution of human immune cells including DC and that human adaptive immunity is clearly reconstituted in these mice (Traggiai et al. 2004). Another group also reported a significant improvement of human immune cell reconstitution with the newborn NOD/SCID/IL2 receptor γ chain^{null} mouse (Ishikawa et al. 2005). More recently, it was reported that efficient HIV-1 replication and CD4 depletion were shown in HSC-engrafted RAG-2/γ_c^{null} mice after HIV-1 inoculation (Baenziger et al. 2006; Berges et al. 2006; Zhang et al. 2007). We also confirmed that the engraftment procedure in newborn NOG mice with human CD34 created an HIV-1-susceptible mouse, and these mice (NOG-hCD34 mouse) after R5 or X4 HIV-1 inoculation possessed a level of viremia similar to that in HIV-1-infected patients. The NOG-hCD34 mice were also susceptible to infection with Epstein-Barr virus (EBV), which is known to be a human-specific γ herpes virus and to infect B cells. Several weeks after EBV inoculation, we could detect EBV DNA in the peripheral blood and spleen cells and could isolate EBV$^+$ lymphoblastoid cells from these mice (Koyanagi, unpublished observations).

3 Application of NOG Mice for HTLV Infection

3.1 HTLV

Human T-lymphotropic virus type I (HTLV-I) is another human retrovirus that causes adult T-cell leukemia (ATL)/ATL lymphoma and can also be involved in certain demyelinating diseases, tropical spastic paraparesis (TSP)/HTLV-I-associated myelopathy (HAM) (Hinuma et al. 1981; Osame et al. 1986). The malignant cells, mostly CD4$^+$ T cells, associated with all phases of ATL express very high levels of IL-2Rα (CD25) (Uchiyama et al. 1985). The median survival duration of all patients with aggressive ATL is about 13 months, and overall survival at 2 years is estimated to be about 30% (Taylor and Matsuoka 2005).

3.2 Model of HTLV-I-Induced Tumorigenicity

SCID mice have been utilized in a study on the pathomechanism and therapeutic strategy of ATL. Imada et al. examined the tumorigenicity of HTLV-I-infected cell lines in an in vivo cell proliferation model using SCID mice. They found that 4 of 11 HTLV-I-infected cell lines were capable of proliferating in SCID mice after intraperitoneal inoculation. Interestingly, it was shown that some HTLV-I-infected and IL-2-dependent cell lines could be successfully engrafted in SCID mice (Imada et al. 1995). The expression of IL-2 mRNA was not detected in these cell lines growing either in vivo or in vitro. No HTLV-I viral structural proteins were detected in three of four transplantable cell lines proliferating in vivo. Peripheral blood T cells immortalized by introduction of *tax* gene of HTLV-I were found to have no tumorigenic potential in SCID mice. Although these systems were useful, two major drawbacks, namely, the long period of time required for tumor formation and the limitation of its use to certain cell lines, appear to hinder wider use of this animal model. These problems have now been overcome through development of the NOG mouse.

PBMC from patients with ATL were inoculated either intraperitoneally into the abdominal region or subcutaneously in the postauricular region of unconditional NOG mice. All mice developed clinical signs of near death, such as piloerection, weight loss, and cachexia, 6-8 weeks after inoculation of ATL cells in addition to enlargement of lymph nodes, spleen, lungs, and liver, whereas no tumors were found in the postauricular region or abdominal cavity where primary ATL cells were inoculated (Dewan et al. 2006). There was no difference in respect to the successful engraftment of ATL cells either intraperitoneally or subcutaneously inoculated into NOG mice. Histologic analysis of ATL-bearing mice showed massive infiltration of leukemic cells in various organs of NOG mice that were efficiently expressing human CD4 and CD25 molecules. A higher level of IL-2R (CD25) expression was observed on the surface of malignant cells associated with all stages of ATL as well as ATL cells infiltrated into various organs of patients. Thus, results from this model indicated successful engraftment and massive infiltration of primary ATL cells in various organs of NOG mice, like leukemia but without producing tumors at the sites of inoculation.

3.3 Evaluation of Anti-ATL Compounds

The various chemotherapies so far developed have not significantly increased the survival of patients with ATL (Taylor and Matsuoka 2005). Given the disappointing results using conventional chemotherapy, new approaches for the treatment of ATL are required. HTLV-I-infected cell lines derived from a leukemic cell clone and primary ATL cells failed to express significant amounts of Tax and other viral proteins, suggesting that the expression of viral proteins is not always necessary for leukemic proliferation at the late stage of the disease. However, HTLV-I-infected cell lines and leukemic cells from patients with ATL display constitutive NF-κB binding activity and increased degradation of a specific inhibitor, IκBα (Mori et al. 1999). NF-κB activation has been connected with multiple processes of oncogenesis, including control of apoptosis, cell cycle, differentiation, and cell migration; therefore, inhibition of NF-κB was suggested to be a useful strategy for cancer therapy. Despite the diversity in clinical manifestations of ATL, strong and constitutive NF-κB activation was reported to be a unique and common characteristic of ATL cells (Mori et al. 1999). Thus, the indispensability of NF-κB for the maintenance of the malignant phenotype of HTLV-I provides a possible molecular target for ATL therapy. To study the effect of an NF-κB inhibitor, ritonavir, on ATL, we injected primary ATL cells from 10 patients subcutaneously into the postauricular region of NOG mice. Beginning 1 day after inoculation, mice were treated with either RPMI-1640 (as control) or ritonavir intraperitoneally daily for 30 days, followed by observation for another 30 days without any treatment. ATL cell inoculation promoted the development of piloerection, weight loss, and cachexia in addition to enlargement of lymph nodes, spleen, lungs, and liver in all control mice 2 months after inoculation. In contrast, ritonavir-treated mice were apparently healthy and had almost no enlargement of these organs. Clinical evaluation of organ invasion 2 months after injection of primary ATL cells showed that ritonavir treatment inhibited their infiltration into lymph nodes, spleen, lungs, and liver (Dewan et al. 2006). Seven of 10 patient samples injected in mice treated with ritonavir presented substantial inhibition of organ invasion, and 2 showed partial inhibition, whereas 1 sample failed to do so. In contrast, all control mice showed formation of new lymph nodes and infiltration with ATL cells into various organs. Organ infiltration of primary ATL cells was analyzed and evaluated by pathological staining and immunostaining of CD4 and CD25. We also performed similar experiments with HTLV-I-infected cell line cells [ED-40515(−), SLB-1, MT-1, TL-Oml, Hut-102, MT-2, and MT-4], using Bay 11-7082, which is another selective inhibitor of TNF-induced phosphorylation of IκBα without affecting the constitutive activation of IκBα phosphorylation, eventually resulting in decreased NF-κB and decreased expression of adhesion molecules (Dewan et al. 2003). Essentially, the same results were obtained in the studies using BAY11-7082 and the ED-40515(−) cell line. Together, these data indicate that NF-κB antagonists significantly inhibit ATL cell growth and infiltration in various organs of NOG mice.

Our NOG ATL model presents many features 6-8 weeks after inoculation of ATL cells such as the clinical signs observed in patients with ATL. Two clinical types, acute and chronic, carry very different prognoses. However, no difference in cell growth, surface phenotype, and NF-κB activity is observed in primary leukemic cells from patients with acute- and chronic-type ATL. Therefore, the same characteristics of freshly isolated ATL cells with acute- and chronic-type were observed in the NOG mouse. Thus, it represents a novel model to evaluate tissue toxicity and the efficacy of therapeutic agents directed toward the treatment of ATL (Fig. 4).

4 Concluding Remarks on Human Retrovirus Infection Model

The humanized mouse model should provide a midway position between the laboratory and clinical studies. From a relevant animal model for HIV and HTLV infections, immunological as well as virological aspects of the disease could be

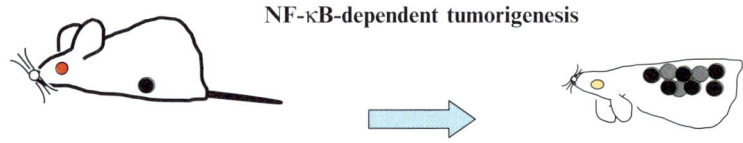

Clinical signs in NOG ATL model
Development to CD4$^+$CD25$^+$ lymphoma
Invasion of leukemic cells to multiple organs
Cachexia
No expression of HTLV-I proteins

Fig. 4 NOG ATL model. Model of HTLV-I-induced tumorigenicity is created with NOG mice. HTLV$^+$ T cells as well as primary ATL leukemic cells induce the identical clinical signs observed in patients with ATL

investigated, and such an animal model could be used to examine the disease process and efficacy of antiviral compounds or vaccines. Significant progress in creating refined mouse strains has been achieved. We have now humanized mice, animals that circulate human blood. However, a mouse is not human. Therefore, there are still many limitations for examination of the human defense system. It is expected that the developments of technology in experimental animals and embryology, including methods using human ES or some other progenitor cells, will open the new evolution of humanized animals.

Acknowledgements This work was supported by grants from the Ministry of Health, Labor and Welfare, and the Ministry of Education, Culture, Sports, Science and Technology of Japan.

References

Aldrovandi GM, Feuer G, Gao L, Jamieson B, Kristeva M, Chen IS, Zack JA (1993) The SCID-hu mouse as a model for HIV-1 infection. Nature 363:732-736

Baenziger S, Tussiwand R, Schlaepfer E, Mazzucchelli L, Heikenwalder M, Kurrer MO, Behnke S, Frey J, Oxenius A, Joller H, Aguzzi A, Manz MG, Speck RF (2006) Disseminated and sustained HIV infection in CD34+ cord blood cell-transplanted Rag2–/–gamma c–/– mice. Proc Natl Acad Sci USA 103:15951-15956

Berges BK, Wheat WH, Palmer BE, Connick E, Akkina R (2006) HIV-1 infection and CD4 T cell depletion in the humanized Rag2–/–gamma c–/– (RAG-hu) mouse model. Retrovirology 3:76

Blunt T, Finnie NJ, Taccioli GE, Smith GC, Demengeot J, Gottlieb TM, Mizuta R, Varghese AJ, Alt FW, Jeggo PA et al. (1995) Defective DNA-dependent protein kinase activity is linked to VDJ recombination and DNA repair defects associated with the murine scid mutation. Cell 80:813-823

Bonyhadi ML, Rabin L, Salimi S, Brown DA, Kosek J, McCune JM, Kaneshima (1993) HIV induces thymus depletion in vivo. Nature 363:728-732

Bosma GC, Custer RP, Bosma MJ (1983) A severe combined immunodeficiency mutation in the mouse. Nature 301:527-530

Boubnov NV, Weaver DT (1995) scid cells are deficient in Ku and replication protein A phosphorylation by the DNA-dependent protein kinase. Mol Cell Biol 15:5700-5706

Delhem N, F Hadida, G Gorochov, F Carpentier, JP de Cavel, JF Andreani, B Autran, JY Cesbron (1998) Primary Th1 cell immunization against HIVgp160 in SCID-hu mice coengrafted with peripheral blood lymphocytes and skin. J Immunol 161:2060-2069

Dewan MZ, Terashima K, Taruishi M, Hasegawa H, Ito M, Tanaka Y, Mori N, Sata T, Koyanagi Y, Maeda M, Kubuki Y, Okayama A, Fujii M, Yamamoto N (2003) Rapid tumor formation of human T-cell leukemia virus type 1-infected cell lines in novel NOD-SCID/γc^{null} mice: suppression by an inhibitor against NF-kappaB. J Virol 77:5286-5294

Dewan MZ, Uchihara JN, Terashima K, Honda M, Sata T, Ito M, Fujii N, Uozumi K, Tsukasaki K, Tomonaga M, Kubuki Y, Okayama A, Toi M, Mori N, Yamamoto N (2006) Efficient intervention of growth and infiltration of primary adult T-cell leukemia cells by an HIV protease inhibitor, ritonavir. Blood 107:716-724

Gorantla S, Santos K, Meyer V, Dewhurst S, Bowers WJ, Federoff HJ, Gendelman HE, Poluektova L (2005) Human dendritic cells transduced with herpes simplex virus amplicons encoding human immunodeficiency virus type 1 (HIV-1) gp120 elicit adaptive immune responses from human cells engrafted into NOD/SCID mice and confer partial protection against HIV-1 challenge. J Virol 79:2124-2132

Hartley O, Gaertner H, Wilken J, Thompson D, Fish R, Ramos A, Pastore C, Dufour B, Cerini F, Melotti A, Heveker N, Picard L, Alizon M, Mosier D, Kent S, Offord R (2004) Medicinal chemistry applied to a synthetic protein: development of highly potent HIV entry inhibitors. Proc Natl Acad Sci USA 101:16460-16465

Hinuma Y, Nagata K, Hanaoka M, Nakai M, Matsumoto T, Kinoshita KI, Shirakawa S, Miyoshi I (1981) Adult T-cell leukemia: antigen in an ATL cell line and detection of antibodies to the antigen in human sera. Proc Natl Acad Sci USA 78:6476-6480

Hiramatsu H, Nishikomori R, Heike T, Ito M, Kobayashi K, Katamura K, Nakahata T (2003) Complete reconstitution of human lymphocytes from cord blood CD34+ cells using the NOD/SCID/γ_c^{null} mice model. Blood 102:873-880

Ifversen P, Borrebaeck CA (1996) SCID-hu-PBL: a model for making human antibodies? Semin Immunol 8:243-248

Imada K, Takaori-Kondo A, Akagi T, Shimotohno K, Sugamura K, Hattori T, Yamabe H, Okuma M, Uchiyama T (1995) Tumorigenicity of human T-cell leukemia virus type I-infected cell lines in severe combined immunodeficient mice and characterization of the cells proliferating in vivo. Blood 86:2350-2357

Ishikawa F, Yasukawa M, Lyons B, Yoshida S, Miyamoto T, Yoshimoto G, Watanabe T, Akashi K, Shultz LD, Harada M (2005) Development of functional human blood and immune systems in γ chain^{null} mice. Blood 106:1565-1573

Ito M, Hiramatsu H, Kobayashi K, Suzue K, Kawahata M, Hioki K, Ueyama Y, Koyanagi Y, Sugamura K, Tsuji K, Heike T, Nakahata T (2002) NOD/SCID/γ_c^{null} mouse: an excellent recipient mouse model for engraftment of human cells. Blood 100:3175-3182

Jamieson BD, Zack JA (1999) Murine models for HIV disease. AIDS 13 Suppl A:S5-11

Kaneshima H, Shih CC, Namikawa R, Rabin L, Outzen H, Machado SG, McCune JM (1991) Human immunodeficiency virus infection of human lymph nodes in the SCID/hu mouse. Proc Natl Acad Sci USA 88:4523-4527

Kaneshima H, Su L, Bonyhadi ML, Connor RI, Ho DD, McCune JM (1994) Rapid-high, syncytium-inducing isolates of human immunodeficiency virus type 1 induce cytopathicity in the human thymus of the SCID-hu mouse. J Virol 68:8188-8192

Kawano Y, Tanaka Y, Misawa N, Tanaka R, Kira JI, Kimura T, Fukushi M, Sano K, Goto T, Nakai M, Kobayashi T, Yamamoto N, Koyanagi Y (1997) Mutational analysis of human immunodeficiency virus type 1 (HIV-1) accessory genes: requirement of a site in the nef gene for HIV-1 replication in activated CD4+ T cells in vitro and in vivo. J Virol 1:8456-8466

Kirchgessner CU, Patil CK, Evans JW, Cuomo CA, Fried LM, Carter T, Oettinger MA, Brown JM (1995) DNA-dependent kinase (p350) as a candidate gene for the murine SCID defect. Science 267:1178-1183

Koyanagi Y, Tanaka Y, Kira J, Ito M, Hioki K, Misawa N, Kawano Y, Yamasaki K, Tanaka R, Suzuki Y, Ueyama Y, Terada E, Tanaka T, Miyasaka M, Kobayashi T, Kumazawa Y, Yamamoto N (1997a) Primary human immunodeficiency virus type 1 viremia and central nervous system invasion in a novel hu-PBL-immunodeficient mouse strain. J Virol 71:2417-2424

Koyanagi Y, Tanaka Y, Tanaka R, Misawa N, Kawano Y, Tanaka T, Miyasaka M, Ito M, Ueyama Y, Yamamoto N (1997b) High levels of viremia in hu-PBL-NOD-scid with HIV-1 infection. Leukemia 11 Suppl 3:109-112

Lu W, LC Arraes, WT Ferreira, J-M Andrieu (2004) Therapeutic dendritic-cell vaccine for chronic HIV-1 infection. Nat Med 10:1359-1365

Lu W, X Wu, Y Lu, W Guo, JM Andrieu (2003) Therapeutic dendritic-cell vaccine for simian AIDS. Nat Med 9:27-32

McCune JM, Kaneshima H, Krowka J, Namikawa R, Outzen H, Peault B, Rabin L, Shih CC, Yee E, Lieberman M et al. (1988) The SCID-hu mouse: murine model for the analysis of human hematolymphoid differentiation and function. Science 241:1632-1639

McCune JM, Kaneshima H, Krowka J, Namikawa R, Outzen H, Peault B, Rabin L, Shih CC, Yee E, Lieberman M et al. (1991) The SCID-hu mouse: a small animal model for HIV infection and pathogenesis. Annu Rev Immunol 9:399-429

McCune JM, Namikawa R, Shih CC, Rabin L, Kaneshima H (1990) Suppression of HIV infection in AZT-treated SCID/hu mice. Science 247:564-566

Miller RD, Hogg J, Ozaki JH, Gell D, Jackson SP, Riblet R (1995) Gene for the catalytic subunit of mouse DNA-dependent protein kinase maps to the scid locus. Proc Natl Acad Sci USA 92:10792-10795

Miura Y, Misawa N, Kawano Y, Okada H, Inagaki Y, Yamamoto N, Ito M, Yagita H, Okumura K, Mizusawa H, Koyanagi Y (2003) Tumor necrosis factor-related apoptosis-inducing ligand induces neuronal death in a murine model of HIV central nervous system infection. Proc Natl Acad Sci USA 100:2777-2782

Miura Y, Misawa N, Maeda N, Inagaki Y, Tanaka Y, Ito M, Kayagaki N, Yamamoto N, Yagita H, Mizusawa H, Koyanagi Y (2001) Critical contribution of TNF-related apoptosis-inducing ligand (TRAIL) to apoptosis of human $CD4^+$ T cells in HIV-1-infected hu-PBL-NOD-SCID mice. J Exp Med 193:651-659

Mori N, Fujii M, Ikeda S, Yamada Y, Tomonaga M, Ballard DW, Yamamoto N (1999) Constitutive activation of NF-kappaB in primary adult T-cell leukemia cells. Blood 93:2360-2368

Mosier DE, Gulizia RJ, Baird SM, Wilson DB (1988) Transfer of a functional human immune system to mice with severe combined immunodeficiency. Nature 335:256-259

Mosier DE, Gulizia RJ, Baird SM, Wilson DB, Spector DH, Spector SA (1991) Human immunodeficiency virus infection of human-PBL-SCID mice. Science 251:791-794

Mosier DE, Gulizia RJ, MacIsaac PD, Torbett BE, Levy JA (1993) Rapid loss of CD4+ T cells in human-PBL-SCID mice by noncytopathic HIV isolates. Science 260:689-692

Nakata H, Maeda K, Miyakawa T, Shibayama S, Matsuo M, Takaoka Y, Ito M, Koyanagi Y, Mitsuya H (2005) Potent Anti-R5-human immunodeficiency virus type 1 effects of a CCR5 antagonist, AK602/ONO4128/GW873140, in a novel human peripheral blood mononuclear cell nonobese diabetic-SCID, interleukin 2 receptor γ-chain-knocked-out AIDS mouse model. J Virol 79: 2087-2096

Namikawa R, Kaneshima H, Lieberman M, Weissman IL, McCune JM (1988) Infection of the SCID-hu mouse by HIV-1. Science 242:1684-1686

Nimura F, Zhang LF, Okuma K, Tanaka R, Sunakawa H, Yamamoto N, Tanaka Y. (2006) Cross-linking cell surface chemokine receptors leads to isolation, activation, and differentiation of monocytes into potent dendritic cells. Exp Biol Med 231:431-443

Nonoyama S, Smith FO, Ochs HD (1993) Specific antibody production to a recall or a neoantigen by SCID mice reconstituted with human peripheral blood lymphocytes. J Immunol. 151:3894-3901

Osame M, Usuku K, Izumo S, Ijichi N, Amitani H, Igata A, Matsumoto M, Tara M (1986) HTLV-I associated myelopathy, a new clinical entity. Lancet 1:1031-1032

Pastore C, Picchio GR, Galimi F, Fish R, Hartley O, Offord RE, Mosier DE (2003) Two mechanisms for human immunodeficiency virus type 1 inhibition by N-terminal modifications of RANTES. Antimicrob Agents Chemother 47:509-517

Peterson SR, Kurimasa A, Oshimura M, Dynan WS, Bradbury EM, Chen DJ (1995) Loss of the catalytic subunit of the DNA-dependent protein kinase in DNA double-strand-break-repair mutant mammalian cells. Proc Natl Acad Sci USA 92:3171-3174

Rabin L, Hincenbergs M, Moreno MB, Warren S, Linquist V, Datema R, Charpiot B, Seifert J, Kaneshima H, McCune J (1996) Use of standardized SCID/hu Thy/Liv mouse model for preclinical efficacy testing of anti-human immunodeficiency virus type 1 compounds. Antimicrob Agents Chemother 40:755-762

Ruxrungtham K, Boone E, Ford H Jr, Driscoll JS, Davey RT Jr, Lane HC (1996) Potent activity of 2'-beta-fluoro-2',3'-dideoxyadenosine against human immunodeficiency virus type 1 infection in hu-PBL-SCID mice. Antimicrob Agents Chemother 40:2369-2374

Sandhu JS, Gorczynski R, Shpitz B, Gallinger S, Nguyen HP, Hozumi N (1995) A human model of xenogeneic graft-versus-host disease in SCID mice engrafted with human peripheral blood lymphocytes. Transplantation 60:179-184

Sandhu JS, Shpitz B, Gallinger S, Hozumi N (1994) Human primary immune response in SCID mice engrafted with human peripheral blood lymphocytes. J Immunol 152:3806-3813

Santini S M, C Lapenta, M Logozzi, S Parlato, M Spada, T Di Pucchio, F Belardelli (2000) Type I interferon as a powerful adjuvant for monocyte-derived dendritic cell development and activity in vitro and in Hu-PBL-SCID mice. J Exp Med 191:1777-1788

Stanley SK, McCune JM, Kaneshima H, Justement JS, Sullivan M, Boone E, Baseler M, Adelsberger J, Bonyhadi M, Orenstein J et al. (1993) Human immunodeficiency virus infection of the human thymus and disruption of the thymic microenvironment in the SCID-hu mouse. J Exp Med 178:1151-1163

Strizki JM, Xu S, Wagner NE, Wojcik L, Liu J, Hou Y, Endres M, Palani A, Shapiro S, Clader JW, Greenlee WJ, Tagat JR, McCombie S, Cox K, Fawzi AB, Chou CC, Pugliese-Sivo C, Davies L, Moreno ME, Ho DD, Trkola A, Stoddart CA, Moore JP, Reyes GR, Baroudy BM (2001) SCH-C (SCH 351125), an orally bioavailable, small molecule antagonist of the chemokine receptor CCR5, is a potent inhibitor of HIV-1 infection in vitro and in vivo. Proc Natl Acad Sci USA 98:12718-12723

Tary-Lehmann M and Saxon A (1992) Human mature T cells that are anergic in vivo prevail in SCID mice reconstituted with human peripheral blood. J Exp Med 175:503-516

Tary-Lehmann M, Saxon A, Lehmann PV (1995) The human immune system in hu-PBL-SCID mice. Immunol Today 16:529-533

Taylor GP, Matsuoka M (2005) Natural history of adult T-cell leukemia/lymphoma and approaches to therapy. Oncogene 24:6047-6057

Traggiai E, Chicha L, Mazzucchelli L, Bronz L, Piffaretti JC, Lanzavecchia A, Manz MG (2004) Development of a human adaptive immune system in cord blood cell-transplanted mice. Science 304:104-107

Uchiyama T, Hori T, Tsudo M, Wano Y, Umadome H, Tamori S, Yodoi J, Maeda M, Sawami H, Uchino H (1985) Interleukin-2 receptor (Tac antigen) expressed on adult T cell leukemia cells. J Clin Invest 76:446-453

Vandekerckhove BA, Namikawa R, Bacchetta R, Roncarolo MG (1992) Human hematopoietic cells and thymic epithelial cells induce tolerance via different mechanisms in the SCID-hu mouse thymus. J Exp Med 175:1033-1043

Watanabe S, Terashima K, Ohta S, Horibata S, Yajima M, Shiozawa Y, Dewan MZ, Yu Z, Ito M, Morio T, Shimizu N, Honda M, Yamamoto N (2007) Hematopoietic stem cell-engrafted NOD/SCID/IL2Rgamma null mice develop human lymphoid systems and induce long-lasting HIV-1 infection with specific humoral immune responses. Blood 109:212-218

Yoshida A, Tanaka R, Kodama A, Yamamoto N, Ansari AA, Tanaka Y (2005) Identification of HIV-1 epitopes that induce the synthesis of a R5 HIV-1 suppression factor by human CD4+ T cells isolated from HIV-1 immunized hu-PBL SCID mice. Clin Dev Immunol 12:235-242

Yoshida A, Tanaka R, Murakam M, Takahashi T, Koyanagi Y, Nakamura M, Ito M, Yamamoto N, Tanaka Y (2003) Induction of protective immune responses against R5 HIV-1 infection in the hu-PBL-SCID mice by intra-splenic immunization with HIV-1-pulsed dendritic cells: possible involvement of a novel factor of human $CD4^+$ T cell origin. J Virol 77:8719-8728

Zhang L, Kovalev GI, Su L (2007) HIV-1 infection and pathogenesis in a novel humanized mouse model. Blood 109:2978-2981

Functional and Phenotypic Characterization of the Humanized BLT Mouse Model

A. K. Wege, M. W. Melkus, P. W. Denton, J. D. Estes, and J. V. Garcia(✉)

1 Introduction 150
   1.1 Definition of the Problem Addressed 150
   1.2 Human/Mouse Xenograft Models to Study Human Hematopoiesis 151
   1.3 Generation of the Bone Marrow/Liver/Thymus Humanized Mouse Model............ 152
2 Human Hematopoietic Reconstitution in BLT Mice 153
   2.1 Peripheral Blood Reconstitution of BLT Mice with Human
       Hematopoietic Cells 153
   2.2 Systemic Reconstitution of BLT Mice with Human Hematopoietic Cells 154
3 T Cell Development in Humanized BLT Mice.. 154
   3.1 Human Thymopoeisis in the BLT Humanized Mouse Model 154
   3.2 Human MHC-Restricted T Cell Response to EBV in BLT Mice 156
   3.3 In Vivo T Cell Response to Toxic Shock Syndrome Toxin-1 in BLT Mice 158
4 Functional Characterization of Human Dendritic Cells in Humanized BLT Mice 158
   4.1 Human Dendritic Cells in Humanized BLT Mice 158
   4.2 In Vivo Analysis of the Human DC Response to TSST-1 in BLT Mice 159
5 Summary and Conclusions 161
References 163

Abstract T cells play a central role in the development of immune responses. Patients lacking T cells because of genetic defects such as DiGeorge or Nezelof syndromes and patients infected with the human immunodeficiency virus are highly susceptible to infections and cancers. The lack of adequate in vivo models of T cell neogenesis have hindered the development and clinical implementation of effective therapeutic modalities aimed at treating these and other clinically important maladies. Transplantation of severe combined immunodeficient (SCID) mice with human hematopoietic stem cells results in long-term engraftment and systemic reconstitution with human progenitor, B, and myeloid cells, but curiously, human T cells are rarely present in any tissue. While the implantation of SCID mice with human fetal thymus and liver (SCID-hu thy/liv mice) allows the

---

J. Victor Garcia
Department of Internal Medicine, Division of Infectious Diseases Y9.206, University of Texas Southwestern Medical Center at Dallas, 5323 Harry Hines Blvd., Dallas, TX 75390-9113, USA
victor.garcia@utsouthwestern.edu

development of abundant thymocytes that are localized in the human organoid implant, there is minimal systemic repopulation with human T cells. However, we have recently shown that transplantation of autologous human hematopoietic fetal liver CD34$^+$ cells into the nonobese diabetic (NOD)/SCID mouse background previously implanted with fetal thymic and liver tissues results in long-term, systemic human T cell homeostasis. In addition to human T cells, these mice have systemic repopulation with human B cells, monocytes/macrophages, and dendritic cells (DC). Importantly, in these mice the T cells developed in the human thymic implant are capable of being activated by human antigen-presenting cells and mount potent human MHC-restricted T cell immune responses.

Abbreviations APC: antigen-presenting cell; BLT: Bone marrow/liver/thymus; DC: dendritic cell; EBV: Epstein-Barr virus; EGFP: enhanced green fluorescent protein; HIV: human immunodeficiency virus; IL-7: interleukin 7; LCL: lymphoblastoid cell line; MHC: major histocompatibility complex; SCID: severe combined immunodeficiency; TCR: T cell receptor; TSST-1: toxic shock syndrome toxin 1

1 Introduction

1.1 Definition of the Problem Addressed

Lack of T cells in humans due to primary or secondary immune deficiencies as in patients with genetic defects such as DiGeorge or Nezelof syndromes or in patients infected with the human immunodeficiency virus (HIV) results in heightened susceptibility to a variety of infections and cancers (Markert et al. 1999; Shearer et al. 1978). The development of effective therapeutic modalities to treat these types of diseases has been hindered, in part, by the lack of adequate in vivo models of human T cell neogenesis. Small animal models of human hematopoiesis have been extensively used as surrogates to study a variety of important aspects of human transplantation, immune function, bacterial and viral infection, tumorigenesis, gene transfer, and in vivo repopulating potential of embryonic and somatic stem cells (Akkina et al. 1994; Aldrovandi et al. 1993; Islas-Ohlmayer et al. 2004; Lapidot et al. 1992, 1997; 1992; McCune et al. 1988; Palucka et al. 2003). Despite their enormous success in a multitude of applications, each of these systems has significant limitations that curtail its usefulness. The development of improved systems with long-term systemic reconstitution with a full complement of human hematopoietic cells is obtained and would provide unique opportunities to study a multitude of timely and relevant issues that cannot be addressed with current models.

1.2 Human/Mouse Xenograft Models to Study Human Hematopoiesis

Immunodeficient mouse models, in general, have been extensively used to study the in vivo function of human hematopoietic stem cell engraftment and reconstitution, important aspects of gene transfer and gene therapy, the replication of human viruses, and the role of human cytokines in hematopoietic homeostasis and, more recently, to study the ontogeny and function of different components of the human immune system, including dendritic cells (DC) (Cravens et al. 2005; Greiner et al. 1998; Islas-Ohlmayer et al. 2004; Lapidot et al. 1992, 1997; Palucka et al. 2003). Perhaps the two in vivo systems most widely used to study different aspects of human hematopoeisis are one in which human $CD34^+$ cells are used to repopulate SCID mice (Greiner et al. 1998) and another in which a small piece of fetal liver or fetal bone is coimplanted with fetal thymic tissue underneath the skin or kidney capsule (SCID-hu thy/liv) (McCune ct al. 1988). In the first example, transplantation of preconditioned SCID mice with human $CD34^+$ cells results in relatively low levels of systemic reconstitution of different mouse tissues with human hematopoietic cells (Greiner et al. 1998). However, one limiting aspect of this system is the fact that B cells are the most abundant cell type in these mice, accounting for up to 90% human reconstitution in all tissues, and virtually no human T cells can be detected in any of the mouse hematopoietic organs including the mouse thymus. In addition, SCID mice show relatively low levels of reconstitution with human cells in general.

One important improvement to the original transplantation model using SCID mice was the use of other immunodeficient strains like NOD/SCID mice, as recipients for transplantation. Transplantation of NOD/SCID mice with human $CD34^+$ cells results in dramatically higher levels of systemic repopulation with human cells. However, despite the higher levels of human reconstitution, T cells generally fail to develop in this system (Greiner et al. 1998; Islas-Ohlmayer et al. 2004). Nevertheless, this model has proven to be extremely useful to study a variety of important aspects of hematopoeisis, the pathogenesis of human-specific virus infection, and the ontogeny and function of the human immune system in vivo (Bente et al. 2005; Cravens et al. 2005; Islas-Ohlmayer et al. 2004; Miyoshi et al. 1999; Palucka et al. 2003).

In contrast to the lack of human T cells in SCID or NOD/SCID mice transplanted with purified $CD34^+$ cells, in the SCID-hu thy/liv mice there is an abundance of human thymocytes. However, virtually all human cells are confined to the thymic organoid that develops after implantation (Aldrovandi et al. 1993; McCune et al. 1988; Vandekerckhove et al. 1991), except for the spleen, where low levels (<1%) of human T cells (and rarely human B cells) can be found. Thus, SCID-hu thy/liv mice do not have significant systemic repopulation with human T cells and are virtually devoid of human B cells, monocytes/macrophages, and dendritic cells. Despite these limitations the SCID-hu thy/liv system has been instrumental as a surrogate model to study hematopoietic stem cell function, thymopoeisis, HIV infection, and pathogenesis (Akkina et al. 1994; Aldrovandi et al. 1993; Amado et al. 1999;

Bonyhadi et al. 1993; Brooks et al. 2001; Jenkins et al. 1998; Kitchen et al. 2000; Kollmann et al. 1995; Napolitano et al. 2003; Okamoto et al. 2002; Su 1997).

Multilineage human reconstitution including T cells in newly developed strains of immunodeficient mice transplanted as neonates with human $CD34^+$ cells from different sources has been reported (Gimeno et al. 2004; Ishikawa et al. 2005; Shultz et al. 2005; Traggiai et al. 2004). The requirements for T cell development in these different strains vary, but for the most part, injection of neonatal mice seems to result in production of T cells, whereas transplantation of adult mice of the same strains results in low to undetectable levels of human T cells that in one instance could be increased by repeated administration of a human IL-7 analog (Shultz et al. 2005). The fact that T cells only arise in mice transplanted as neonates indicates the need for a developing system conducive to expansion of the T cell compartment. The single common denominator in all these recently described systems is the fact that the selection of the human T cell repertoire is believed to take place in the context of the mouse MHC in the mouse thymus (Ishikawa et al. 2005; Ito et al. 2002; Shultz et al. 2005; Traggiai et al. 2004; Yahata et al. 2002). In these mouse models, positive selection presumably occurs when human thymocyte TCRs interact with mouse MHC molecules loaded with mouse self-peptide on murine thymic epithelial cells.

1.3 Generation of the Bone Marrow/Liver/Thymus Humanized Mouse Model

There is a great interest in generating humanized mice in which human T cells develop in the context of a human thymic environment in order to understand human T cell ontogeny and human MHC-restricted T cell responses in vivo. When autologous or allogeneic human $CD34^+$ progenitor cells are provided with an appropriate thymic microenvironment, such as being injected directly into the thymic organoid that develops in the SCID-hu thy/liv model, they can mature into naive single-positive human T cells (Akkina et al. 1994; An et al. 1997). We therefore asked whether $CD34^+$ cells introduced via bone marrow transplantation could systemically reconstitute the mouse and sustain thymopoiesis in the implanted human thymic tissue. In order to maximize the likelihood of seeding the transplanted thymus with human T cell progenitors originating from the human stem cells present in the mouse bone marrow, we used NOD/SCID mice instead of SCID mice, because NOD/SCID mice support significantly higher levels of reconstitution after transplantation with human $CD34^+$ cells because of lower endogenous mouse NK cell activity (Greiner et al. 1998).

Allogeneic cord blood $CD34^+$ cells were originally used in this system because of their accessibility and high in vivo repopulating potential (Vormoor et al. 1994). However, regardless of their source, transplantation with allogeneic $CD34^+$ cells failed to sustain the implanted thymic tissue and resulted in clearance of the bone marrow graft in this model. Therefore, autologous fetal liver $CD34^+$ cells were used to determine whether they could contribute to the overall levels of human reconstitution when transplanted after implantation of MHC-matched human fetal thymus and liver.

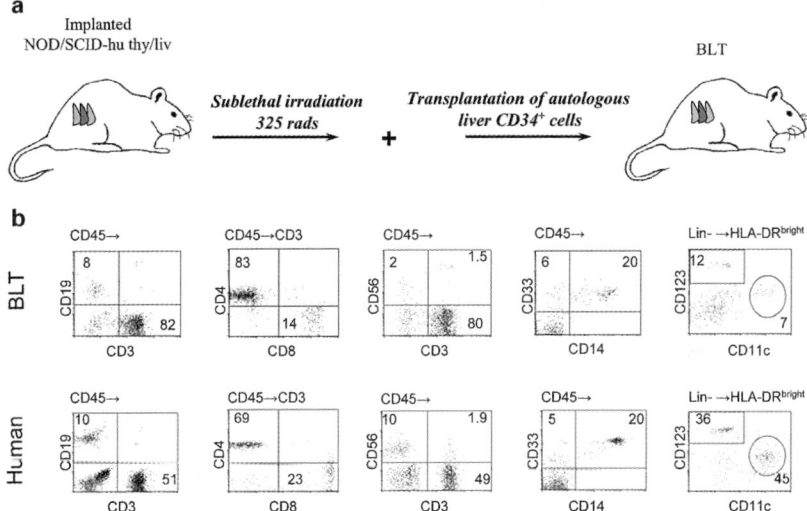

Fig. 1 Reconstitution of NOD/SCID-hu thy/liv mice transplanted with autologous human hematopoietic stem cells. **a** Schematic illustration of the different steps involved in generating NOD/SCID-hu BLT mice. **b** flow cytometry analysis of peripheral blood from one representative BLT mouse and a healthy human control. Samples were analyzed by gating on CD45 and then for the specific antigen (i.e., CD45→CD3→CD4 or CD8, CD45→CD19). The percentages of the individual human cell subsets in each sample are indicated by the *numbers* in each of the quadrants. Human DC were defined as lineage-negative and HLA-DR^{bright} cells expressing either CD123 or CD11c (i.e., lin⁻→HLA-DR^{bright}→CD123 or CD11c)

In essence, NOD/SCID mice were implanted with human fetal thymic and liver tissues, preconditioned with a sublethal dose of gamma radiation, and transplanted with autologous human CD34⁺ cells obtained from a portion of the same fetal liver used for the implant (Fig. 1a). To emphasize the importance of the reconstitution of the mouse bone marrow with human hematopoietic stem cells to repopulate the previously implanted fetal liver and thymic tissue, we have adopted the designation NOD/SCID-hu BLT mice (or BLT mice) to distinguish them from the SCID-hu thy/liv model.

2 Human Hematopoietic Reconstitution in BLT Mice

2.1 Peripheral Blood Reconstitution of BLT Mice with Human Hematopoietic Cells

Human cells represent a minor proportion of cells in the blood of mice coimplanted with human thymus and liver alone (Akkina et al. 1994; Aldrovandi et al. 1993; Amado et al. 1999; McCune et al. 1988, 1989). However, previously implanted NOD/SCID-hu thy/liv mice that received a bone marrow transplant with autologous CD34⁺ cells had readily detectable levels of human hematopoietic cells in their peripheral blood (Fig. 1b) (Melkus et al. 2006). When peripheral blood cells were characterized for different human hematopoietic lineages, B cells, monocyte/macrophages, dendritic cells, and specifically T cells were present in all reconstituted mice by 8 weeks and sustained for up to 26 weeks after transplant (Fig. 1b) (Melkus et al. 2006). These results demonstrated that the peripheral blood of bone marrow-transplanted NOD/SCID-hu thy/liv mice with purified autologous CD34$^+$ cells results in significant levels of multilineage peripheral blood reconstitution with human hematopoietic cells including T cells.

2.2 Systemic Reconstitution of BLT Mice with Human Hematopoietic Cells

Since human hematopoietic cells were readily found in the peripheral blood of BLT mice, the presence of human cells in primary and secondary lymphoid tissues was determined. Overall levels of reconstitution with human cells in different mouse tissues are summarized in Fig. 2a. In contrast to SCID-hu thy/liv mice, in BLT mice bone marrow, lymph nodes, spleen, thymic organoid, liver, lung, and gut demonstrated substantial levels of human CD45$^+$ cells (Fig. 2a). Therefore in BLT mice primary and secondary lymphoid organs as well as no-lymphoid tissues important in immune regulation and mucosal immunity are reconstituted with human hematopoietic cells. On analysis of these tissues, we noted the presence of human T cells and their subpopulations, B cells (Fig. 2b), monocyte/macrophages, and dendritic cells (Melkus et al. 2006). With respect to the human T cell subsets present in the peripheral blood, the mean of CD4$^+$ cells was 70% and the mean of CD8$^+$ T cells was 20% (Fig. 2c). The differentiation state of the human T cells in the peripheral blood of BLT mice was analyzed by determining the expression of CD45RA and CD27. Approximately 62% of the T cells in the periphery exhibited a naive phenotype, with 32% exhibiting a central memory phenotype (Fig. 2c, bottom). As expected for animals kept under sterile conditions, BLT mice had a higher percentage of naive T cells when compared to healthy human controls. In summary, these data demonstrated that BLT mice develop a remarkable state of sustained systemic multilineage reconstitution with human hematopoietic cells that can persist for months after transplantation.

3 T Cell Development in Humanized BLT Mice

3.1 Human Thymopoeisis in the BLT Humanized Mouse Model

One of the distinctive features of BLT mice is the fact that human T cells are generated in the context of a human thymus (Melkus et al. 2006). The relative proportion of double- and single-positive thymocytes of the human thymic organoid in BLT

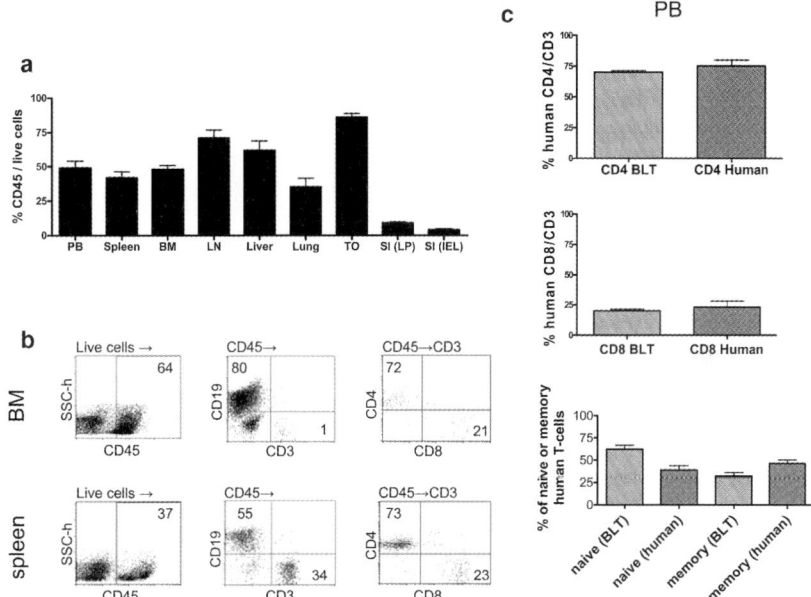

Fig. 2 Analysis of human hematopoietic reconstitution of different tissues in BLT mice. **a** Systemic human reconstitution in peripheral blood (*PB*; $n=20$), spleen ($n=24$), bone marrow (*BM*; $n=24$), lymph nodes (*LN*; $n=14$), liver ($n=5$), lung ($n=7$), thymic organoid (*TO*; $n=24$), small intestine lamina propria [*SI (LP)*; $n=24$], and small intestine intraepithelial lymphocytes [*SI (IEL)*; $n=24$] of BLT mice. Human reconstitution in each tissue was determined by flow cytometry gating on live cells expressing human CD45. **b** Analysis of the B and T cell subsets in the bone marrow (*BM*) and spleen of a representative BLT mouse. Samples were analyzed by gating on CD45 and then for the specific antigen (i.e., CD45→CD3→CD4 or CD8, CD45→CD19). **c** Comparison of the percentage of CD4⁺ and CD8⁺ T cell subsets and naive vs. memory cells in the peripheral blood of BLT mice ($n=55$) vs. healthy human controls ($n=4$)

mice is similar to that observed in human fetal and child thymi. Despite the radiation treatment received by the implanted thymic organoid in the BLT mice, these animals maintained relatively constant levels of $CD4^+CD8^+$ T cells in the human thymic organoid over their life span, demonstrating long-term sustained thymopoiesis (Fig. 3a, b). In addition, human T cells in BLT mice were found to express a diverse repertoire of Vβ T cell receptors (TCR) (Fig. 3d).

Progenitor seeding of the implanted human thy/liv organoid from transplanted bone marrow stem cell progenitors and their contribution in sustaining thymopoiesis by generating de novo T cells in the BLT mouse model was demonstrated by using a lentivirus-based vector expressing enhanced green fluorescent protein (EGFP). In essence, autologous fetal liver CD34⁺ cells were first transduced with an EGFP-expressing lentivirus-based vector. Transduced cells were then used to reconstitute NOD/SCID-hu thy/liv mice (Melkus et al. 2006). Human cells expressing CD45 and EGFP were found in bone marrow, spleen, and thymic organoid (all the tissues examined). Analysis of

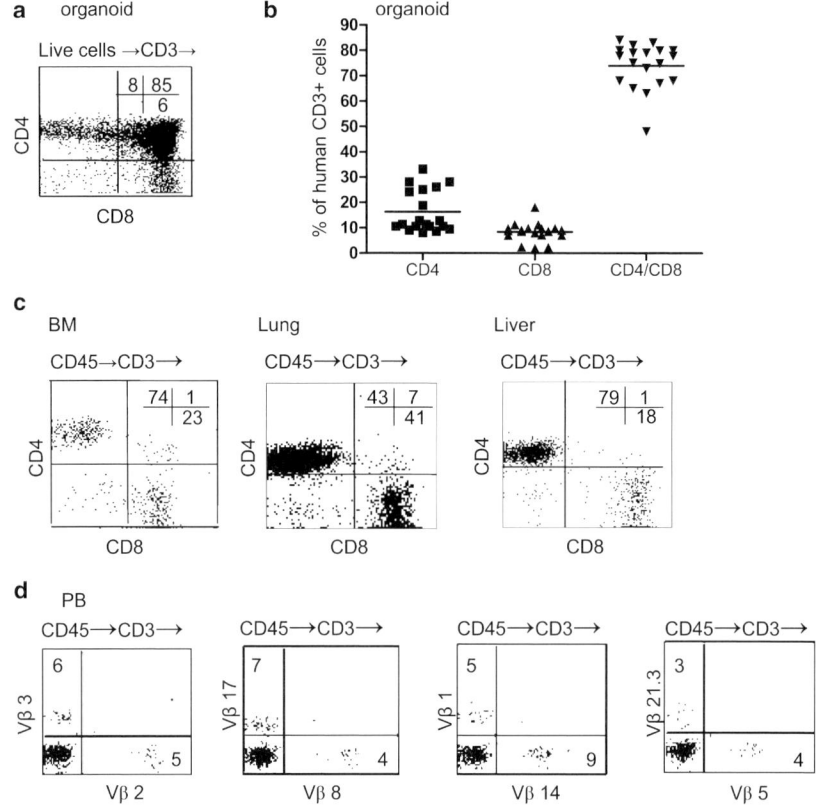

Fig. 3 Human thymopoiesis in BLT mice. **a** Flow cytometry analysis for CD4 and CD8 expression in the thymic organoid of a representative BLT mouse. **b** Analysis of thymic reconstitution of single- and double-positive thymocytes (*lines* represent medians). **c** Analysis of T cell subsets in BM, lung, and liver of one representative BLT mouse. **d** Analysis of selected T cell receptors in peripheral blood human T cells of BLT mice

EGFP expression in different hematopoietic lineages, as well as T cell subsets, further demonstrated that the transplanted CD34+ cells contributed to the overall levels of engraftment and reconstitution of all human lineages examined. Specifically, they contributed to the seeding of the thymic organoid with human T cell progenitors, thus contributing to de novo thymopoiesis and overall T cell homeostasis.

3.2 Human MHC-Restricted T Cell Response to EBV in BLT Mice

In the BLT mouse model human T cells develop properly in an autologous human thymic environment. Therefore it was important to determine whether these human T cells recognize antigens in the context of human MHC to generate specific T cell responses. This was accomplished with a clinically relevant approach, namely, the development of a T cell response to Epstein-Barr Virus (EBV). In parallel to the implantation of the fetal liver/thymic tissue and the isolation of autologous CD34⁺ cells, fetal liver B cells were used to establish autologous lymphoblastoid cell lines (LCLs) to serve as antigen-presenting cells (APC) (Fig. 4a). Autologous BLT mice were then infected with EBV. Peripheral blood T cells from infected animals demonstrated a dramatic increase in the percentage of CD45RA⁻CD27⁺ central memory T cells (Melkus et al. 2006). This was reflective of the pattern of expansion of T cells in the peripheral blood of patients during acute EBV infection (Roos et al. 2000).

When human T cells isolated from different organs were cocultured with the previously established autologous antigen-presenting LCLs, the human T cells produced significant levels of γ-interferon (Fig. 4b). γ-Interferon production was inhibited when LCLs were pretreated with anti-human MHC class I and/or class II

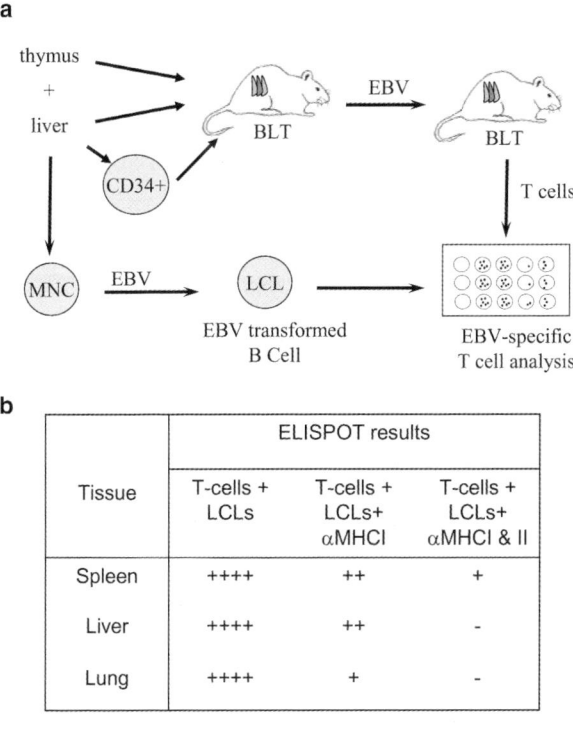

Fig. 4 Characterization of the human MHC-restricted immune response to EBV in BLT mice. **a** Diagram showing the different steps used to induce an EBV-specific immune response and analyzing for EBV-specific γ-interferon production. **b** Results of ELISPOT analysis for the presence of EBV-specific T cells producing γ-interferon from spleen, liver. and lung. Also shown are the inhibitory effects of anti-human MHC I alone or in combination with anti-human MHC II antibodies. Symbols: −, <10 spots; +, 10-20 spots; ++, 30-60 spots; ++++, 100-180 spots per 10^6 human T cells

antibodies, demonstrating that human T cells developed in BLT mice produce human MHC-restricted responses to EBV and highlighting the dynamic interaction between human T and B cells in this model.

3.3 In Vivo T Cell Response to Toxic Shock Syndrome Toxin-1 in BLT Mice

T cell responses in vivo are governed by their interaction with APC. As demonstrated by the experiments described above, in BLT mice human B cells can present viral antigens to human T cells, resulting in a specific cellular immune response. It was then important to demonstrate that human APC are also capable of inducing in vivo a specific human T cell response. For this purpose, toxic shock syndrome toxin-1 (TSST-1) was chosen as a stimulus because of its clinical relevance in human disease and because the immunological response to TSST-1 in humans has been well characterized (Dinges et al. 2000). TSST-1 specifically activates and induces TCR Vβ2$^+$ T cells to proliferate through cross-linking of the TCR on T cells and MHC II on APC (Makida et al. 1996). Consistent with the specificity of TSST-1 for T cells expressing the Vβ2 TCR, a dramatic expansion of this specific subset of human T cells was observed in the periphery of BLT mice. Furthermore, an increase in the number of human Vβ2$^+$ T cells was also observed in all animals after TSST-1 administration (Fig. 5a).

Massive cytokine production resulting from exposure to TSST-1 is a key factor in the pathogenesis of toxic shock syndrome in humans (Kum et al. 2001). The plasma levels of human cytokines in control and TSST-1-treated BLT mice showed significant increases in the systemic levels of human INF-γ, IL-10, IL-6, IL-8, and TNFα (Fig. 5b). This cytokine profile in response to TSST-1 in BLT mice resembles that seen in humans (Kum et al. 2001), further demonstrating that human T cells within BLT mice are capable of exerting effector function after TCR stimulation by APC.

4 Functional Characterization of Human Dendritic Cells in Humanized BLT Mice

4.1 Human Dendritic Cells in Humanized BLT Mice

Dendritic cells (DC) play important roles in health and disease and are generated in the bone marrow from CD34$^+$ cells. NOD/SCID mice reconstituted with human CD34$^+$ cells have been shown to have a full repertoire of human DC. In fact, the ontogeny of human DC in NOD/SCID mice transplanted with human CD34$^+$ cells

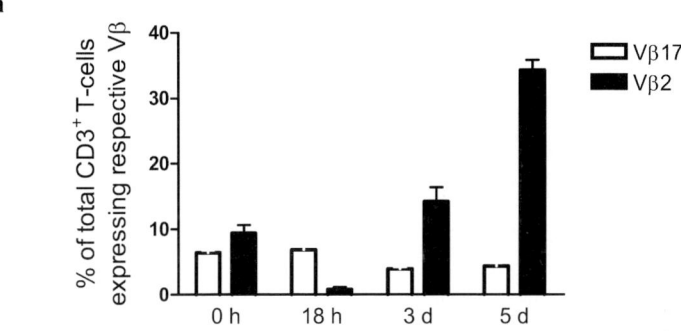

Cytokine	Control (n=4)	TSST-1	
		1h after (n=4)	18h after (n=6)
IFN γ	50.1 +/- 25.1	106.1 +/- 35.9	479.9 +/- 143.0
Il-10	24.4 +/- 1.9	85.1 +/- 26.8	257.6 +/- 71.8
TNF α	13.4 +/- 1.2	197.7 +/- 109.9	23.5 +/- 2.9
Il-8	31.0 +/- 2.6	46.1 +/- 6.2	128.1 +/- 36.1
Il-6	33.8 +/- 13.1	107.2 +/-18.2	335.3 +/- 92.7

Fig. 5 Innate immune response to TSST-1 in BLT mice. **a** Increase in the percentage of Vβ2⁺ T cells at different times after TSST-1 injection. Note the lack of response by the Vβ17⁺ human T cells that served as internal control for these experiments. **b** Cytokine production (±SEA) in response to TSST-1 in the plasma of BLT mice before and 1 h and 18 h after TSST-1 administration

was shown to be parallel what is seen in humans in all aspects (Cravens et al. 2005). For their analysis in humanized BLT mice, DC were defined as lineage-negative, HLA-DR^{bright} CD11c⁺ or CD123⁺. In BLT mice, DC are systemically distributed. DC were found in peripheral blood, bone marrow, lymph nodes, thymic organoid, spleen, liver, lung, and gut.

4.2 In Vivo Analysis of the Human DC Response to TSST-1 in BLT Mice

Superantigens induce T cell responses by bridging the T cell receptor and the MHC II molecules expressed on the surface of APC (Dinges et al. 2000; Karp et al. 1990). Whereas the T cell response has been extensively studied in humans, the response by APC has not been so carefully described. The extent and the kinetics of the phenotypic changes to human DC in response to systemic administration of TSST-1 in vivo regarding activation and maturation antigens in BLT mice provided some highly relevant and novel information. Before TSST-1 administration, CD123+ and CD11c+ DC in bone marrow of BLT mice have an immature resting phenotype characterized by either low levels or the absence of human CD40, CD80, CD86, and CD83 surface expression (Melkus et al. 2006). Similarly, CD123+ and CD11c+ DC in the spleen of unmanipulated BLT mice have an immature resting phenotype (Fig. 6).

TSST-1 administration did not significantly alter the phenotype of human CD123+ DC in the bone marrow or spleen of BLT mice. TSST-1 administration did not alter the phenotype of bone marrow CD11c+ DC, either. In contrast, a dramatic (but transient) upregulation of activation and maturation markers was observed in the CD11c+ DC present in the spleen of BLT mice treated with TSST-1 (Fig. 6). Administration of TSST-1 to a set of transplanted mice with a full complement of human DC but devoid of human T cells (because they were not implanted with human fetal thymic and liver tissue before transplant) did not result in upregulation of DC activation and/or maturation markers, demonstrating

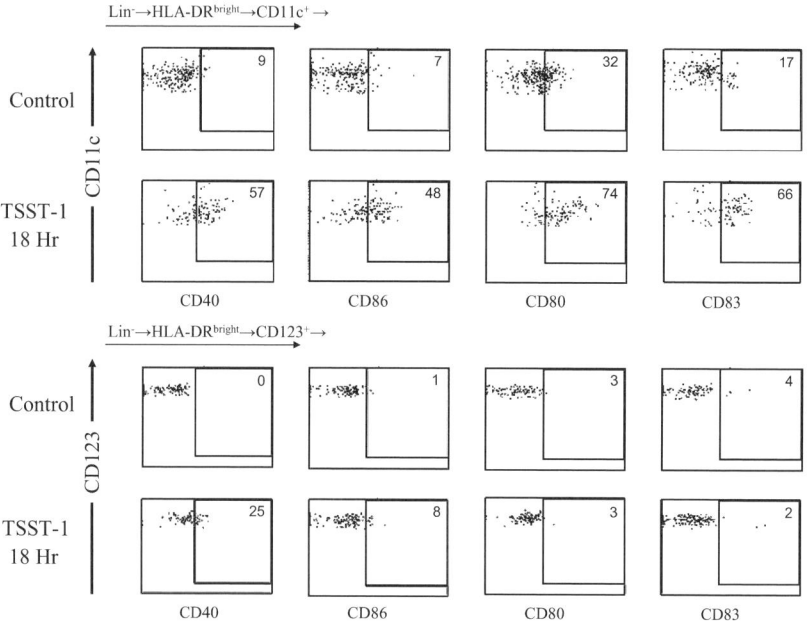

Fig. 6 In vivo response of human dendritic cells isolated from the spleen of BLT mice to TSST-1. *Top* Upregulation of activation and maturation markers on CD11c+ DC. ***Bottom*** Analysis of activation and maturation marker expression on CD123+ DC. Human DC were defined as lineage-negative and HLA-DR^{bright} cells expressing either CD123 or CD11c (i.e., lin⁻→HLA-DR^{bright}→ CD123+ or CD11c+)

that in combination with the T cells present in the BLT mice, human CD11c$^+$ DC respond to TSST-1 by upregulating activation and maturation markers in spleen but not bone marrow.

5 Summary and Conclusions

Hematological abnormalities that result in immunodeficiency are generally associated with a multitude of complications and relatively poor prognosis. For example, DiGeorge syndrome is a rare congenital disorder in which the thymus fails to develop (Markert et al. 1999). Thus, DiGeorge syndrome patients develop all aspects of a functional immune system except that they are T cell deficient, making them susceptible to numerous opportunistic infections. Transplantation of allogeneic thymic tissue into patients with DiGeorge syndrome has been shown to restore T cell production and restore significant immune function (Markert et al. 1999). This advance was made possible by knowledge derived from in vivo experimentation in mice without other available systems that could serve to bridge the transition between experimentation in mice and the clinical implementation of this knowledge in humans (Weissman and Shizuru 1999). During the past two decades, the development and implementation of experimental models to study human hematopoietic and immune system dysfunction have been greatly facilitated by the availability of several immune-deficient mouse strains capable of accepting human grafts (Greiner et al. 1998; Ishikawa et al. 2005; Ito et al. 2002; Shultz et al. 2005; Traggiai et al. 2004; Yahata et al. 2002). Whereas none of these models fully recapitulates all aspects of human hematopoeisis, individually they all have served in the study of a variety of aspects regarding hematopoietic stem cell function as well as immune function.

Transplantation of SCID or NOD/SCID mice with human hematopoietic stem cells recapitulates several important aspects of DiGeorge and Nezelof syndromes, namely, thymic atrophy and the development of multilineage human hematopoeisis in the absence of T cells (Markert et al. 1999; Shearer et al. 1978). The molecular basis for the T cell lineage restriction in these mice is currently unknown. In SCID-hu thy/liv mice there is long-term sustained thymopoiesis in the graft without systemic peripheral reconstitution, preventing a wider utilization of this otherwise very useful model (Aldrovandi et al. 1993; McCune et al. 1988; Vandekerckhove et al. 1991).

This is in contrast to SCID or NOD/SCID mice transplanted with human hematopoietic stem cells. In transplanted mice there is systemic reconstitution with hematopoietic cells, but T cells are not generated (Greiner et al. 1998; Islas-Ohlmayer et al. 2004). The absence of human T cells in transplanted SCID or NOD/SCID mice was hypothesized as being due to the lack of an appropriate microenvironment for human T cell development. To test this hypothesis we implanted NOD/SCID mice with human fetal thymus and liver, preconditioned them with a sublethal dose of radiation, and transplanted them with autologous CD34⁺ human hematopoietic cells (BLT mice). Transplantation of human CD34⁺ cells into previously implanted mice resulted in systemic repopulation with multilineage human hematopoietic cells including T cells (Melkus et al. 2006).

Phenotypic analysis of human cells in the peripheral blood of BLT mice matches that of cells from normal humans. In peripheral blood, spleen, lung, liver, and bone marrow of BLT mice human single-positive CD4⁺ and CD8⁺ T cells, B cells, monocytes, macrophages, and both CD123⁺ and CD11c⁺ DC are present, indicating multilineage hematopoietic reconstitution in the different primary and secondary lymphoid organs as well as important tertiary lymphoid tissues. In the implanted human thymic organoid double-negative, double-positive, and single-positive human thymocytes are present, in contrast to secondary lymphoid organs and other nonlymphoid organs like the gut, liver, and lung, where for the most part only single-positive T cells are present.

In contrast to other humanized animal models, in BLT mice thymocytes undergo positive and negative selection in the human thymic organoid in the context of autologous MHC restriction and not in the mouse thymus. In fact, no human T cells are detected in the mouse thymic tissue of fully reconstituted BLT mice. These observations indicate a preferential trafficking of human hematopoietic cells into the human thymic tissue in BLT mice.

By infecting BLT mice with EBV, a specific MHC I- and MHC II-restricted human T cell immune response was demonstrated to occur. Analyses of secondary lymphoid tissue like the spleen, sites of immune regulation like the liver, and mucosal sites like the lung all showed significant T cell responses. Studies of EBV infection in animals lacking human T cells has shown tumor development (Islas-Ohlmayer et al. 2004). This was not noted in BLT mice, perhaps the result of protection by a T cell immune response.

Superantigens such as TSST-1 have been shown to be the causative agents of a variety of human diseases (Dinges et al. 2000). TSST-1 triggers an excessive cellular immune response that can lead to systemic release of cytokines, the expansion of Vβ2⁺ T cells, and lethal toxic shock. In addition, it has been suggested that bacterial and viral superantigens in general may play a major role in generating a break in immune tolerance and increased epitope spreading in autoimmune diseases (Soos et al. 2002). A powerful aspect of the BLT model was demonstrated by showing that the developed human immune system can be dissected to determine the role of not only human T cells but also of human DC during an immune response.

The BLT model allowed for genetically identical mice to be generated with a complete human immune system or, in nonimplanted mice, an immune system devoid of human T cells. This attribute facilitates studies of how human T cells interact with other immune cells, specifically DC, and how these interactions might result in preventing or inducing disease in vivo. Two important aspects of the in vivo response to TSST-1 are worth emphasizing. First, no significant phenotypic changes were noted in CD123⁺ DC in BLT mice. Second, administration of TSST-1 to animals devoid of human T cells (i.e., reconstituted with human CD34⁺ cells but not implanted with human thy/liv tissues) did not result in the production of detectable levels of human cytokines or any changes in the phenotype of human DC. These results demonstrate the potential of this system to investigate the cellular interactions of the human innate immune response to superantigens and to evaluate novel therapeutic modalities aimed at preventing the devastating effects of toxic shock syndrome and other superantigens in general.

In summary, BLT mice represent a novel model of human hematopoiesis that incorporates the best attributes of two well-established systems: the SCID-hu thy/liv mouse and the NOD/SCID-hu bone marrow transplant mouse. BLT mice develop de novo T cells within a human thymic environment and generate a systemic human hematopoietic system, which is maintained long term in all tissues examined and is capable of mounting a human MHC-restricted T cell immune response. Practical models of human hematopoiesis are essential to bridge the gaps in our understanding of human immune system response and maturation. No animal model will completely recapitulate all aspects of human hematopoiesis or the human immune system. However, new and improved models that closely recapitulate key aspects of human hematopoiesis will serve to provide insight into fundamental aspects of human stem cell engraftment and reconstitution, immune system development, new strategies to study human pathogenesis, and novel therapeutic approaches to alleviate or cure human diseases.

Acknowledgements We would like to thank the members of our laboratories who contributed to the development of the BLT model. We would also like to thank Drs. M. Manz, L. Shultz, O. Sharma, J. Turpin, and Ashley Haase for their continued support and encouragement during the development and implementation of the BLT model. This work was supported in part by the Department of Obstetrics and Gynecology's Tissue Procurement Facility of the University of Texas Southwestern Medical Center at Dallas (NIH Grant HD011149), Grants CA-82055, AI-71940 and AI-39416 (J.V.G.), and Training Grant T32 AI-07421 (J.D.E.).

References

Akkina, R. K., Rosenblatt, J. D., Campbell, A. G., Chen, I. S., and Zack, J. A. (1994). Blood 84, 1393-8.
Aldrovandi, G. M., Feuer, G., Gao, L., Jamieson, B., Kristeva, M., Chen, I. S., and Zack, J. A. (1993). Nature 363, 732-6.
Amado, R. G., Jamieson, B. D., Cortado, R., Cole, S. W., and Zack, J. A. (1999). J Virol 73, 6361-9.
An, D. S., Koyanagi, Y., Zhao, J. Q., Akkina, R., Bristol, G., Yamamoto, N., Zack, J. A., and Chen, I. S. (1997). J Virol 71, 1397-404.
Bente, D. A., Melkus, M. W., Garcia, J. V., and Rico-Hesse, R. (2005) J. Virol 79, 1397-1399.
Bonyhadi, M. L., Rabin, L., Salimi, S., Brown, D. A., Kosek, J., McCune, J. M., and Kaneshima, H. (1993). Nature 363, 728-32.
Brooks, D. G., Kitchen, S. G., Kitchen, C. M., Scripture-Adams, D. D., and Zack, J. A. (2001). Nat Med 7, 459-64.
Cravens, P. D., Melkus, M. W., Padgett-Thomas, A., Islas-Ohlmayer, M., Del, P. M. M., and Garcia, J. V. (2005). Stem Cells 23, 264-78.

Dinges, M. M., Orwin, P. M., and Schlievert, P. M. (2000). Clin Microbiol Rev 13, 16-34.
Gimeno, R., Weijer, K., Voordouw, A., Uittenbogaart, C. H., Legrand, N., Alves, N. L., Wijnands, E., Blom, B., and Spits, H. (2004). Blood 104, 3886-93.
Greiner, D. L., Hesselton, R. A., and Shultz, L. D. (1998). Stem Cells 16, 166-77.
Ishikawa, F., Yasukawa, M., Lyons, B., Yoshida, S., Miyamoto, T., Yoshimoto, G., Watanabe, T., Akashi, K., Shultz, L. D., and Harada, M. (2005). Blood 106, 1565-73.
Islas-Ohlmayer, M., Padgett-Thomas, A., Domiati-Saad, R., Melkus, M. W., Cravens, P. D., Martin Mdel, P., Netto, G., and Garcia, J. V. (2004). J Virol 78, 13891-900.
Ito, M., Hiramatsu, H., Kobayashi, K., Suzue, K., Kawahata, M., Hioki, K., Ueyama, Y., Koyanagi, Y., Sugamura, K., Tsuji, K., Heike, T., and Nakahata, T. (2002). Blood 100, 3175-82.
Jenkins, M., Hanley, M. B., Moreno, M. B., Wieder, E., and McCune, J. M. (1998). Blood 91, 2672-8.
Karp, D. R., Teletski, C. L., Scholl, P., Geha, R., and Long, E. O. (1990). Nature 346, 474-6.
Kitchen, S. G., Killian, S., Giorgi, J. V., and Zack, J. A. (2000). J Virol 74, 2943-8.
Kollmann, T. R., Kim, A., Pettoello-Mantovani, M., Hachamovitch, M., Rubinstein, A., Goldstein, M. M., and Goldstein, H. (1995). J Immunol 154, 907-21.
Kum, W. W., Cameron, S. B., Hung, R. W., Kalyan, S., and Chow, A. W. (2001). Infect Immun 69, 7544-9.
Lapidot, T., Fajerman, Y., and Kollet, O. (1997). J Mol Med 75, 664-73.
Lapidot, T., Pflumio, F., Doedens, M., Murdoch, B., Williams, D. E., and Dick, J. E. (1992). Science 255, 1137-41.
Makida, R., Hofer, M. F., Takase, K., Cambier, J. C., and Leung, D. Y. (1996). Mol Immunol 33, 891-900.
Markert, M. L., Boeck, A., Hale, L. P., Kloster, A. L., McLaughlin, T. M., Batchvarova, M. N., Douek, D. C., Koup, R. A., Kostyu, D. D., Ward, F. E., Rice, H. E., Mahaffey, S. M., Schiff, S. E., Buckley, R. H., and Haynes, B. F. (1999). N Engl J Med 341, 1180-9.
McCune, J. M., Kaneshima, H., Lieberman, M., Weissman, I. L., and Namikawa, R. (1989). Curr Top Microbiol Immunol 152, 183-93.
McCune, J. M., Namikawa, R., Kaneshima, H., Shultz, L. D., Lieberman, M., and Weissman, I. L. (1988). Science 241, 1632-9.
Melkus, M. W., Estes, J. D., Padgett-Thomas, A., Gatlin, J., Denton, P. W., Othieno, F. A., Wege, A. K., Haase, A. T., and Garcia, J. V. (2006). Nat Med 12, 1316-22.
Miyoshi, H., Smith, K. A., Mosier, D. E., Verma, I. M., and Torbett, B. E. (1999). Science 283, 682-6.
Napolitano, L. A., Stoddart, C. A., Hanley, M. B., Wieder, E., and McCune, J. M. (2003). J Immunol 171, 645-54.
Okamoto, Y., Douek, D. C., McFarland, R. D., and Koup, R. A. (2002). Blood 99, 2851-8.
Palucka, A. K., Gatlin, J., Blanck, J. P., Melkus, M. W., Clayton, S., Ueno, H., Kraus, E. T., Cravens, P., Bennett, L., Padgett-Thomas, A., Marches, F., Islas-Ohlmayer, M., Garcia, J. V., and Banchereau, J. (2003). Blood 102, 3302-10.
Roos, M. T., van Lier, R. A., Hamann, D., Knol, G. J., Verhoofstad, I., van Baarle, D., Miedema, F., and Schellekens, P. T. (2000). J Infect Dis 182, 451-8.
Shearer, W. T., Wedner, H. J., Strominger, D. B., Kissane, J., and Hong, R. (1978). Pediatrics 61, 619-24.
Shultz, L. D., Lyons, B. L., Burzenski, L. M., Gott, B., Chen, X., Chaleff, S., Kotb, M., Gillies, S. D., King, M., Mangada, J., Greiner, D. L., and Handgretinger, R. (2005). J Immunol 174, 6477-89.
Soos, J. M., Mujtaba, M. G., Schiffenbauer, J., Torres, B. A., and Johnson, H. M. (2002). J Neuroimmunol 123, 30-4.
Su, L. (1997). Rev Med Virol 7, 157-166.
Traggiai, E., Chicha, L., Mazzucchelli, L., Bronz, L., Piffaretti, J. C., Lanzavecchia, A., and Manz, M. G. (2004). Science 304, 104-7.
Vandekerckhove, B. A., Krowka, J. F., McCune, J. M., de Vries, J. E., Spits, H., and Roncarolo, M. G. (1991). J Immunol 146, 4173-9.

Vormoor, J., Lapidot, T., Pflumio, F., Risdon, G., Patterson, B., Broxmeyer, H. E., and Dick, J. E. (1994). Blood Cells 20, 316-20; discussion 320-2.
Weissman, I. L., and Shizuru, J. A. (1999). N Engl J Med 341, 1227-9.
Yahata, T., Ando, K., Nakamura, Y., Ueyama, Y., Shimamura, K., Tamaoki, N., Kato, S., and Hotta, T. (2002). J Immunol 169, 204-9.

Novel Metastasis Models of Human Cancer in NOG Mice

M. Nakamura(✉) and H. Suemizu

1	Introduction	168
2	Animal Experimentation Strategies for Cancer Metastasis	168
	2.1 Hematogenous Metastasis	168
	2.2 Lymphatic Metastasis	169
	2.3 Dissemination	169
3	Liver Metastasis Models of Human Pancreatic Cancer Cells in NOG Mice	169
	3.1 Experimental Liver Metastasis Panel	170
	3.2 Superiority of NOG Mice as a Liver Metastasis Model	171
	3.3 Establishment of a Highly Metastatic Cell Line	172
	3.4 Exploration of Metastasis-Related Genes with Liver Metastasis Models	172
	3.5 Evaluation of Candidate Gene Function with Liver Metastasis Models	173
4	Distant Metastasis Models of Human Melanoma Cells in NOG Mice	173
	4.1 Experimental Distant Metastasis Panel	174
	4.2 Correlation Analysis with Gene Expression and Metastatic Ability	174
5	Future Directions of Metastasis Models of Human Cancer in NOG Mice	175
6	Conclusion	175
References		176

Abstract In this chapter, cancer research using immunodeficient mice, with emphasis on metastasis field studies, is described. The definition of "humanized mice" used for biomedical research is given in the chapter by Nomura et al. in this volume. Briefly, a humanized mouse possesses human cells or tissues and shows, in part, an identical biological function to human beings. However, humanized mice described *in this* chapter may differ slightly from this definition. In research on cancer metastasis, the *in vivo* dynamic state of human cancer cells after transplantation into NOG mice is partially identical to that in human beings. This chapter also describes the superiority of NOG mice over conventional immunodeficient NOD/Shi-*scid* mice in cancer research.

M. Nakamura
Central Institute for Experimental Animals, 1430 Nogawa, Miyamae,
Kawasaki 216-0001, Japan
nakamura.masato@hachioji-hosp.tokai.ac.jp

1 Introduction

Distant metastasis of malignant neoplasms consists of several pathological steps. The neoplastic cells first invade the surrounding stroma. Malignant epithelial tumors, or carcinomas in particular, need to invade beyond the basement membrane, the local barrier for the cancerous cells. Thus, carcinoma occurring *in situ* must invade the subepithelial stroma before it invades the blood and lymphatic vessels. Various complex mechanisms are involved in the first local invasion. When the cancerous cells reach the vasculature, the cells destroy the vascular walls and invade the lumen. The cancerous cells then enter the blood or lymphatic circulation. Most cancerous cells are single, but a few cancerous cellular aggregates are formed. The cancerous cells adhere to or are entrapped in vessels actively or passively. Certain cancerous cells show positive adhesion to the vessel walls through adhesion molecules. The adherent or entrapped cancerous cells destroy or invade the vascular walls and then invade the surrounding stroma with destruction of the cells. The cancerous cells must complete all the above complex steps. *In vivo* analysis using cancer cell lines provides a basis for discussion of invasion mechanisms or local steps in metastatic events. To analyze the complex steps or mechanisms of distant metastases, reliable and reproducible experimental metastatic models must be developed. Xenografts of human malignant neoplastic cells are useful experimental systems for studying metastatic mechanisms of human malignant neoplasms. However, xenograft metastatic experiments are performed in various immune-deficient mice (conventional irradiated mice, *nude* mice, and NOD-*scid* mice). These mouse models have limited potentials, and a limited number of human neoplasms can be established as xenografts. Inoculation of a large number of various established human neoplastic cell lines into immune-deficient mice is necessary to establish distant metastasis in the mice. New concepts, strategies, or tactics are required to analyze mechanisms of human malignant neoplasms under *in situ* or *in vivo* conditions with immune-deficient mice. One possible solution is development of newly designed immune-deficient mice including NOG (NOD/Shi-*scid IL2R* ^{null}) mice. Application of a wide range of human neoplastic cells is expected. Reliable or simple xenotransplantation methods will be essential for quantitative analyses of distant metastasis.

2 Animal Experimentation Strategies for Cancer Metastasis

2.1 Hematogenous Metastasis

Hematogenous metastasis models are rather easy to establish in immune-deficient mice. Conventional inoculation of neoplastic cells into the tail vein can produce initial pulmonary metastatic lesions. These systems are useful for analyzing metastatic mechanisms of human gastrointestinal cancers (colorectal cancer and gastric cancers). Tail vein inoculation, however, results in diverse and irreproducible results in immune-deficient mice. Further simplification is essential to quantitatively analyze hematogenous metastasis. In this sense, intrasplenic or intraportal inoculation is a reliable and reproducible method to produce hepatic metastasis. The route is simpler than other routes of orthotopic inoculation of human resources in immune-deficient mice.

2.2 Lymphatic Metastasis

Lymphatic metastasis models are the most difficult to develop in mice. The lymphatic systems are not developed in mice, especially on node structures. Human cancers first invade the lymphatic vessels and metastasize to the regional or sentinel lymph nodes.

2.3 Dissemination

Dissemination in body cavities is a late or advanced event of human malignant neoplasms. Abdominal dissemination is much easier to develop in mouse models. Abdominal dissemination models can easily be produced by direct inoculation of human malignant neoplasms. Dissemination models give limited information for finding molecular targets of treatment of early-stage cancer.

3 Liver Metastasis Models of Human Pancreatic Cancer Cells in NOG Mice

New strategies for treating metastatic cancer will require the development of appropriate animal models for studying their effectiveness. Several models of liver metastasis using intrasplenic injection of cancer cells have been established and characterized with athymic *nude* mice [2, 11, 13]. Hematogenous metastasis occurs as a consequence of a well-characterized set of sequential events. These types of metastasis models mimic only those events that occur after cells enter the blood vessels. In most of these model systems, more than one million cancer cells are intrasplenically inoculated into mice to generate liver metastases [5, 13–15]. It is unlikely, however, that such a large number of cancer cells would enter the liver at one time via the portal vein and form metastatic foci in patients with pancreatic cancer. Figure 1 is a schematic representation of the liver metastasis model.

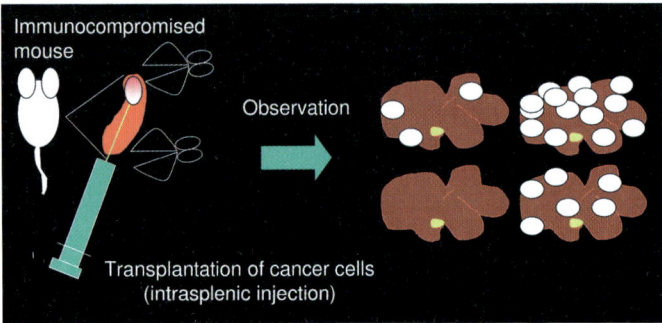

Fig. 1 Schematic representation of experimental protocols. Human cancer cells were transplanted into 7- to 9-week-old male NOG and NOD/Shi-*scid* mice by intrasplenic injection

3.1 Experimental Liver Metastasis Panel

Seven human pancreatic cancer cell lines were examined for their ability to form diverse metastatic foci in the liver of NOD/Shi-*scid* and NOG mice. The incidences of liver metastases in NOG mice were far higher than those in NOD/Shi-*scid* mice (Table 1). Capan-2 and PL45 showed no metastasis when seeded at up to 10^5 cells in both strains, and no BxPC-3 metastasis was observed in NOD/Shi-*scid* mice. The NOD/Shi-*scid* mice model could detect liver metastasis only for the AsPC-1 cell line when inoculated with more than 10^3 cells. In contrast, when NOG mice were inoculated with only 10^2 MIA PaCa-2, AsPC-1, and PANC-1 cells, liver metastasis was evident in 71.4%–37.5% of the mice. The incidence of metastasis of Capan-1 or BxPC-3 cells was greatly reduced when fewer numbers of cells were inoculated into NOG mice. In contrast, apparent metastases were evident in more than 50% of NOG mice inoculated with 10^2 AsPC-1 or MIA PaCa-2 cells. Thus, the occurrence of metastatic lesions in the livers of NOG mice was dose dependent, reproducible, and apparent over a wide range of logarithmic values. The metastatic potential of the different cell lines can be ranked as follows: MIA PaCa-2>AsPC-1>PANC-1>Capan-1>BxPC-3>Capan-2 and PL45. The liver metastatic potential of seven individual pancreatic cancer cell lines in NOG mice was quantitated as the T/L score. The ranking of metastatic potential was quantitatively reproduced with T/L score analysis. Six weeks after 10^4 cells were transplanted, the T/L score of each cell line was as follows: MIA PaCa-2, 60.6%; AsPC-1, 48.2%; PANC-1, 26.6%; and Capan-1, 15.6%. The overall surface area of livers inoculated with pancreatic cancer cells was not significantly different from that of normal (noninoculated) NOG mice livers, with the exception of livers inoculated with MIA PaCa-2 cells. Liver metastatic potential of human pancreatic cancer cells in NOG mice is not directly related to

Table 1 Liver metastasis after intrasplenic injection of human pancreatic cancer cells

Cell line	Cell dose (cells/head)	Number of animals with liver metastasis[a] (metastasis/total)		
		NOD-*scid*	NOD-*scid*+AGM[b]	NOG
MIA PaCa–2	1×10^4	0/10 (0.0%)	4/10 (40.0%)	10/10 (100.0%)
	1×10^3	0/7 (0.0%)	ND	6/6 (100.0%)
	1×10^2	0/6 (0/0%)	ND	4/7 (71.4%)
AsPC–1	1×10^4	8/9 (88.9%)	ND	9/9 (100.0%)
	1×10^3	2/8 (25.0%)	7/10 (70.0%)	8/8 (100.0%)
	1×10^2	0/6 (0.0%)	ND	4/7 (57.1%)
PANC–1	1×10^4	0/10 (0.0%)	0/10 (0.0%)	8/8 (100.0%)
	1×10^3	0/6 (0.0%)	ND	6/8 (75.0%)
	1×10^2	0/7 (0.0%)	ND	3/8 (37.5%)
Capan–1	1×10^4	0/10 (0.0%)	2/10 (20.0%)	9/10 (90.0%)
	1×10^3	0/10 (0.0%)	ND	5/10 (50.0%)
	1×10^2	0/8 (0.0%)	ND	0/8 (0.0%)
BxPC–3	1×10^5	0/8 (0.0%)	8/10 (80.0%)	8/8 (100.0%)
	1×10^4	0/8 (0.0%)	ND	1/8 (12.5%)
Capan–2	1×10^5	0/8 (0.0%)	ND	0/8 (0.0%)
	1×10^4	ND	ND	0/10 (0.0%)
PL45	1×10^5	0/8 (0.0%)	ND	0/8 (0.0%)
	1×10^4	ND	ND	0/10 (0.0%)

ND, not done
[a] Liver metastasis was evaluated at 6 weeks after inoculation of 1×10^3, 10^4, and 10^5 cancer cells and at 8 weeks after inoculation of 10^2 cancer cells
[b] AGM indicates anti-asialo GM1 antibody

tumor growth, because significant differences were not observed in the doubling times and subcutaneous tumorigenicity.

3.2 Superiority of NOG Mice as a Liver Metastasis Model

One of the putative factors that make NOG mice more immunodeficient than NOD/Shi-*scid* mice is the complete lack of natural killer (NK) cell activity. The involvement of NK cells in liver metastases was examined by elimination of NK cell activity with anti-asialo GM1 antibody administration. Elimination of NK cells in NOD/Shi-*scid* mice by treatment with anti-asialo GM1 antibody enhances the liver metastatic incidence of the human pancreatic cancer cell lines, which means that NK cell activity has a crucial role in the rejection of xenotransplanted cells. Representative images of liver metastases are shown in Fig. 2. However, for every cell line we tested, a higher incidence of liver metastasis was observed in NOG mice than in NOD/Shi-*scid* mice with antibody treatment. Treatment with anti-asialo GM1 antibody could not create a condition in NOD/Shi-*scid* mice similar to the immunocompromised condition in NOG mice. The efficient liver metastases seen in NOG mice may reflect elimination of other immune factors in addition to T, B, and NK cells.

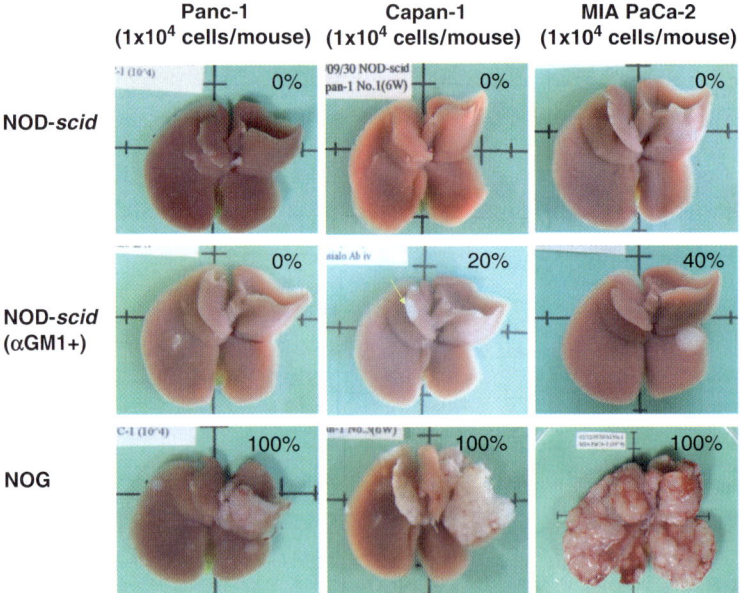

Fig. 2 Comparison of metastatic ability among NOD/Shi-*scid* mice, NOD/Shi-*scid* mice with GM1 treatment, and NOG mice

3.3 Establishment of a Highly Metastatic Cell Line

BxPC-3 cells showed poor metastatic capability in our liver metastasis model using NOG mice, but a few visible tumor foci consistently developed. We isolated cells from these foci, and designated them "LM (Liver-Metastasized)-BxPC-3" cells. After a short period in culture, 1×10^5 LM-BxPC-3 cells were injected into NOG mice. In contrast to the parental cell line, we observed aggressive metastasis at 6 weeks when LM-BxPC-3 cells were injected into NOG mice. Both cell lines had the same genetic profile at four polymorphic microsatellite markers examined (D5S818, D13S317, D7S820, and D16S539) and similar *in vitro* growth potentials.

3.4 Exploration of Metastasis-Related Genes with Liver Metastasis Models

We performed oligonucleotide microarray analysis to identify genes that were differentially expressed in BxPC-3 cells and their derivative, LM-BxPC-3 cells. Among over 47,000 total feature sets present on the chip, when parental cells were compared with LM-BxPC-3 cells, statistically significant differences in gene expression (4-fold or greater) were observed with 45 gene sets. Nine genes had decreased expression levels, whereas 36 genes had increased expression levels in LM-BxPC-3 cells, compared with the parental cell line. Quantitative RT-PCR was carried out to quantitate expression of two underexpressed genes and seven overexpressed genes in the LM-BxPC-3 line. The expression dynamics based on quantitative RT-PCR were in agreement with the GeneChip data. To determine whether the genes identified by expression profiling of BxPC-3 and LM-BxPC-3 cells were associated with liver metastasis in other pancreatic cancer cell lines, we examined gene expression with quantitative RT-PCR and liver metastatic potential with the T/L score of each of the other six pancreatic cancer cell lines. For eight of the nine genes, there was no correlation between gene expression and liver metastasis. However, expression of S100A4 gene strongly correlated with liver metastasis. At the protein level, there was a strong correlation between S100A4 expression and in vivo liver metastatic ability.

3.5 *Evaluation of Candidate Gene Function with Liver Metastasis Models*

The S100A4 gene was identified as a metastasis-associated gene because its expression level correlated closely with liver metastasis. We were interested in the effect of S100A4 overexpression on liver metastatic potential of pancreatic cancer cells. The full-length human S100A4 cDNA was cloned into an expression vector and transfected into BxPC-3 parental cells. Two antibiotic-resistant clones were selected. Clone-G expressed S100A4 at dramatically increased levels, and clone-B showed the same levels as those in parental BxPC-3 cells. Both cell clones were transplanted intrasplenically into NOG mice to evaluate their *in vivo* liver metastatic potential. Metastatic potential was evaluated by the incidence of liver metastasis and T/L score. When NOG mice were inoculated with 10^5 cells, the incidence of liver metastasis in the mice that received either S100A4-transfectants or nontransfectants (BxPC-3 cells) was 100%. The T/L score of each clone was calculated, and the correlation between S100A4 expression and liver metastatic potential was evaluated. Clone-G, which expressed high levels of S100A4 protein, had a significantly higher T/L score than clone-B, which was neomycin resistant but had extremely low levels of S100A4 protein, and nontransfected parental BxPC-3 cells. When NOG mice were inoculated with 10^4 cells, liver metastasis was evident in only 12.5% of NOG mice that received nontransfected parental BxPC-3 cells and was never observed in mice that received the clone-B cells. In contrast, inoculation of 10^4 clone-G cells resulted in apparent metastasis in more than 50% of NOG mice.

4 Distant Metastasis Models of Human Melanoma Cells in NOG Mice

Tumor xenografts in immunodeficient mice (athymic *nude* mice and *scid* mice) are well-established animal models for the study of human cancer. Several human melanoma cell lines were reported to metastasize in athymic *nude* mice [3, 6].

The *scid* mice, which have defective rearrangements of the T-cell receptor and B-cell receptor, resulting in defects of functional T and B cells, were discovered in 1983 [1]. Taylor et al. [19] reported the growth and dissemination of human malignant melanoma cells in *scid* mice and revealed metastatic lesions. However, the metastatic rates were usually low, despite a sufficient number of cells, more than 10^6 cells, being injected.

4.1 Experimental Distant Metastasis Panel

Four human melanoma cell lines were examined for their ability to form distant metastatic foci in NOG and NOD/Shi-*scid* mice. After 6 weeks of intravenous inoculation, the mice were autopsied to evaluate the metastatic foci of the melanoma cells in various organs. All melanoma cell lines showed metastasis in the NOG mice, while no metastatic lesions were observed in the NOD/Shi-*scid* mice. When NOG mice were inoculated with 10^4 A2058, A375, G361, and HMY-1 cells, distant metastasis was evident in 100% (6/6), 89% (8/9), 33% (2/6), and 25% (2/8) of the mice, respectively. Metastatic foci were detected in the liver and lung, while no metastases were identified in other organs such as the brain and digestive organs. A2058, A375, and G361 cells metastasized to the liver in 83% (5/6), 78% (7/9), and 33% (2/6) of the NOG mice, respectively. Distant metastasis directed to lungs was observed in A2058, A375, and HMY-1 cell lines with incidences of 67% (4/6), 56% (5/9), and 25% (2/8). A2058 and A375 cell lines had significantly higher metastasis ratios than G361 and HMY-1 cell lines by Fisher's exact probability test ($p=0.004$).

4.2 Correlation Analysis with Gene Expression and Metastatic Ability

The expression level of the S100A4 protein, a member of the S100 family, has been associated with an increased metastatic capacity of cancer cells [9, 17]. Some studies have suggested a correlation between the expression of S100A4 and the clinical outcome in various tumor types [4, 10, 12]. In other studies, an inverse association between S100A4 and E-cadherin has been revealed [7, 8, 21]. E-cadherin mediates homophilic calcium-dependent cell-cell adhesion [16, 20] and is the prime mediator of melanocyte adhesion to keratinocytes *in vitro* [18]. A2058 and A375 cell lines showed increased gene expression of S100A4 coupled with higher metastatic potentials. Furthermore, E-cadherin gene expression was conversely inhibited in these cell lines. The increased levels of S100A4 combined with inhibited E-cadherin resulted in high metastatic potentials of the human melanoma cell lines *in vivo*.

5 Future Directions of Metastasis Models of Human Cancer in NOG Mice

To date, the NOG mouse is the only *in vivo* model for liver metastasis that can evaluate metastatic potential with an inoculation dose of 10^2 cells. The NOG mice liver metastasis models are reliable, reproducible, and quantitative. This system will be developed and expanded to estimate complex mechanisms of cancer. Especially for malignant epithelial neoplasms or carcinoma, the mechanisms of interactions should be clarified between cancerous epithelial cells and stroma. Metastasizing or floating cancerous cells are considered in single-cell conditions, while no precise conditions are apparent for cellular aggregates. The possibility of active influence of stromal or extracellular matrix cannot be excluded. When we inoculate human cancerous cells mixed with stromal elements or matrices, are the metastatic potentials enhanced or suppressed? Other possibilities for this simple or novel experimental model are considered for cancerous stem cells. Various cellular fractions of the primary cancerous tissues could be inoculated in NOG mice. Cancerous stem cells definitely show metastatic lesions in this model. This means that these methods will condense or isolate cancerous stem cells from primary cancerous tissues.

Hematogenous metastasis models, especially metastasized systemic models, must be optimized for routes and numbers inoculated and the experimental protocols. In systemic metastasis models, we hope to evaluate the metastatic grade without autopsy. Methods for detection and estimation of metastatic lesions have made dramatic advances recently. Noninvasive imaging systems should be a powerful tool for this purpose.

Lymphatic or lymphoid metastasis models have not been completed yet, even though we used the NOG mouse to establish experimental models. Nodal structures are also required for the systems. Therefore, further improvement will be required.

Basically, metastatic models of human cancer reflect the late events of metastasis. The models reflect metastatic events after adhesion or entrapment in the organs. Models must be developed to reflect the initial steps or local invasion mechanisms at the primary lesion of the cancer.

6 Conclusion

NOG mice have no lymphocytes or NK cells and have impaired dendritic cell function. NOG mice are a superior xenotransplantation system for engraftment of human cancer cells compared with NOD/Shi-*scid* mice. A high rate of metastasis in NOG mice inoculated with small numbers of cancer cells (as low as 100 cells in the liver) and a higher level of metastasis in the NOG mouse model than that in NOD/Shi-*scid* mice were demonstrated. These metastasis models using NOG mice are reliable and quantitative, and more closely mimic the *in vivo* conditions in patients with cancer. They will also be valuable tools for exploring new metastasis-related genes with genome-wide gene expression profiles and proteome and metabolome analyses.

References

1. Bosma, G. C., R. P. Custer, and M. J. Bosma. 1983. A severe combined immunodeficiency mutation in the mouse. *Nature* 301:527–530.
2. Bresalier, R. S., S. E. Raper, E. S. Hujanen, and Y. S. Kim. 1987. A new animal model for human colon cancer metastasis. *Int J Cancer* 39:625–630.
3. Cornil, I., S. Man, B. Fernandez, and R. S. Kerbel 1989. Enhanced tumorigenicity, melanogenesis, and metastases of a human malignant melanoma after subdermal implantation in nude mice. *J Natl Cancer Inst* 81:938–944.
4. Gongoll, S., G. Peters, M. Mengel, P. Piso, J. Klempnauer, H. Kreipe, and R. von Wasielewski. 2002. Prognostic significance of calcium-binding protein S100A4 in colorectal cancer. *Gastroenterology* 123:1478–1484.
5. Ikeda, Y., M. Ezaki, I. Hayashi, D. Yasuda, K. Nakayama, and A. Kono. 1990. Establishment and characterization of human pancreatic cancer cell lines in tissue culture and in nude mice. *Jpn J Cancer Res* 81:987–993.
6. Iliopoulos, D., C. Ernst, Z. Steplewski, J. A. Jambrosic, U. Rodeck, M. Herlyn, W. H. Clark, Jr., H. Koprowski, and D. Herlyn. 1989. Inhibition of metastases of a human melanoma xenograft by monoclonal antibody to the GD2/GD3 gangliosides. *J Natl Cancer Inst* 81:440–444.
7. Kimura, K., Y. Endo, Y. Yonemura, C. W. Heizmann, B. W. Schafer, Y. Watanabe, and T. Sasaki. 2000. Clinical significance of S100A4 and E-cadherin-related adhesion molecules in non-small cell lung cancer. *Int J Oncol* 16:1125–1131.
8. Kohya, N., Y. Kitajima, W. Jiao, and K. Miyazaki. 2003. Effects of E-cadherin transfection on gene expression of a gallbladder carcinoma cell line: repression of MTS1/S100A4 gene expression. *Int J Cancer* 104:44–53.
9. Maelandsmo, G. M., E. Hovig, M. Skrede, O. Engebraaten, V. A. Florenes, O. Myklebost, M. Grigorian, E. Lukanidin, K. J. Scanlon, and O. Fodstad. 1996. Reversal of the in vivo metastatic phenotype of human tumor cells by an anti-CAPL (mts1) ribozyme. *Cancer Res* 56:5490–5498.
10. Nakamura, T., T. Ajiki, S. Murao, T. Kamigaki, S. Maeda, Y. Ku, and Y. Kuroda. 2002. Prognostic significance of S100A4 expression in gallbladder cancer. *Int J Oncol* 20:937–941.
11. Nicholson, B. E., H. F. Frierson, M. R. Conaway, J. M. Seraj, M. A. Harding, G. M. Hampton, and D. Theodorescu. 2004. Profiling the evolution of human metastatic bladder cancer. *Cancer Res* 64:7813–7821.
12. Ninomiya, I., T. Ohta, S. Fushida, Y. Endo, T. Hashimoto, M. Yagi, T. Fujimura, G. Nishimura, T. Tani, K. Shimizu, Y. Yonemura, C. W. Heizmann, B. W. Schafer, T. Sasaki, and K. Miwa. 2001. Increased expression of S100A4 and its prognostic significance in esophageal squamous cell carcinoma. *Int J Oncol* 18:715–720.
13. Nishimori, H., T. Yasoshima, F. Hata, R. Denno, Y. Yanai, H. Nomura, H. Tanaka, K. Kamiguchi, N. Sato, and K. Hirata. 2002. A novel nude mouse model of liver metastasis and peritoneal dissemination from the same human pancreatic cancer line. *Pancreas* 24:242–250.
14. Nomura, H., H. Nishimori, T. Yasoshima, F. Hata, H. Tanaka, F. Nakajima, T. Honma, J. Araya, K. Kamiguchi, H. Isomura, N. Sato, R. Denno, and K. Hirata. 2002. A new liver meta-

static and peritoneal dissemination model established from the same human pancreatic cancer cell line: analysis using cDNA macroarray. *Clin Exp Metastasis* 19:391–399.
15. Shishido, T., T. Yasoshima, K. Hirata, R. Denno, M. Mukaiya, H. Ura, K. Yamaguchi, S. Kawaguchi, and N. Sato. 1999. Establishment and characterization of human pancreatic carcinoma lines with a high metastatic potential in the liver of nude mice. *Surg Today* 29:519–525.
16. Takeichi, M. 1991. Cadherin cell adhesion receptors as a morphogenetic regulator. *Science* 251:1451–1455.
17. Takenaga, K., Y. Nakamura, and S. Sakiyama. 1997. Expression of antisense RNA to S100A4 gene encoding an S100-related calcium-binding protein suppresses metastatic potential of high-metastatic Lewis lung carcinoma cells. *Oncogene* 14:331–337.
18. Tang, A., M. S. Eller, M. Hara, M. Yaar, S. Hirohashi, and B. A. Gilchrest. 1994. E-cadherin is the major mediator of human melanocyte adhesion to keratinocytes in vitro. *J Cell Sci* 107 (Pt 4):983–992.
19. Taylor, C. W., T. M. Grogan, M. H. Lopez, S. P. Leong, A. Odeleye, F. J. Feo-Zuppardi, and E. M. Hersh. 1992. Growth and dissemination of human malignant melanoma cells in mice with severe combined immune deficiency. *Lab Invest* 67:130–137.
20. Wheelock, M. J., and P. J. Jensen. 1992. Regulation of keratinocyte intercellular junction organization and epidermal morphogenesis by E-cadherin. *J Cell Biol* 117:415–425.
21. Yonemura, Y., Y. Endou, K. Kimura, S. Fushida, E. Bandou, K. Taniguchi, K. Kinoshita, I. Ninomiya, K. Sugiyama, C. W. Heizmann, B. W. Schafer, and T. Sasaki. 2000. Inverse expression of S100A4 and E-cadherin is associated with metastatic potential in gastric cancer. *Clin Cancer Res* 6:4234–4242.

In Vivo Imaging in Humanized Mice

H. Masuda(✉), H. J. Okano, T. Maruyama, Y. Yoshimura,
H. Okano, and Y. Matsuzaki

1	Introduction	180
2	Clinical Imaging Systems	181
	2.1 Ultrasound Imaging	181
	2.2 Computerized Tomography	181
	2.3 Magnetic Resonance Imaging	182
	2.4 Single-Photon Emission Computed Tomography	183
	2.5 Positron Emission Tomography	184
3	Optical Imaging Systems	184
	3.1 Fluorescence Imaging	185
	3.2 Bioluminescence Imaging	185
	3.3 Comparison of BLI with FLI	186
4	Multimodality Imaging	187
5	A Novel In Vivo Imaging System of the Endometrial Model Mouse	187
	5.1 Lentiviral Introduction of Reporter Genes into Primary Endometrial Cells	189
	5.2 Noninvasive, Real-Time, and Quantitative Assessment of the Reconstructs by BLI	190
	5.3 Discussion of This System	192
6	Conclusion	193
References		194

Abstract The radiological modalities that are currently utilized as critical components in clinical medicine have also been adapted to small-animal imaging, among which are ultrasound imaging, X-ray computerized tomography (CT), magnetic resonance imaging (MRI), positron emission tomography (PET), and single-photon emission computed tomography (SPECT). Optical imaging techniques such as bioluminescence imaging (BLI) and fluorescence imaging (FLI) are approaches that are commonly used in small animals. Longitudinal surveys of living (i.e., nonsacrificed) animal models with these modalities provide some clues for the development of clinical applications. The techniques are absolutely essential

---

H. Masuda
Department of Obstetrics and Gynecology, Keio University School of Medicine,
35 Shinanomachi, Shinjuku-ku, Tokyo 160-8582, Japan
hirotaka@1997.jukuin.keio.ac.jp

for translational research. However, there are currently few tools available with sufficient spatial or temporal resolution ideal for all experimental studies. In this chapter, we provide a rationale and techniques for visualizing target cells in living small animals and an overview of the advantages and limitations of current imaging technology. Finally, we introduce a humanized mouse and a novel in vivo imaging system that we have developed. We also discuss real-time observations of reconstructs and clinical manifestations.

Abbreviations BLI: bioluminescence imaging; CBR luc: click beetle red-emitting luciferase; CCD: charge-coupled device; CT: x-ray computerized tomography; E_2: estradiol; $E_2 + P_4$: treatment with E_2 in combination with P_4; ^{18}FDG: 18-Fluorodeoxyglucose; FLI: fluorescence imaging; fMRI: functional MRI; FMT: fluorescence molecular tomography; γ-rays: gamma rays; GFP: green fluorescent protein; ICI: ICI-182,780; IRES : internal ribosomal entry site; MRI: magnetic resonance imaging; μMRI: micro-MRI; NIR: near-infrared range; NOG: NOD/ SCID/ γ_c^{null}; OVX: Ovariectomized; P_4: progesterone; PET: positron emission tomography; Q-Dot: quantum Dot; RF: radiofrequency; ROI: region of interest; SDECs: singly dispersed endometrial cells; SNR: signal-to-noise ratio; SPECT: single-photon emission computed tomography; $T_{1/2}$: half-time; UBM: ultrasound biomicroscopy; YFP: yellow fluorescent protein; 3D: three-dimensional

1 Introduction

Much of our understanding of various diseases has been obtained by using in vitro culture systems, in which the influences of intact organ structure, circulation, and the immune system have been removed. On the other hand, in vivo studies, especially those using humanized mice, can better mimic the actual physiological condition. In addition, in vivo imaging methods enable longitudinal studies of multiple processes and parameters in individual animals. Novel information about the specific three-dimensional (3D) locations, interaction, and dynamic states can be obtained through in vivo imaging without the necessity of killing the animals.

Accordingly, noninvasive and real time in vivo imaging in animal models including humanized mice holds promise in the provision of biomedical advances. In vivo imaging in humanized mice has become the significant bridge between in vitro basic research and clinical applications. The monitoring of cell growth and detection of dynamic states including metastasis in living animals have paved the way for the development of new drugs and expanded our knowledge of both the pathophysiology and the pathogenesis of each disease studied.

We have developed a mouse model for the study of the human endometrium and endometriosis. Our models are severe immunodeficient female mice transplanted with human endometrial cells. These diseased mice transplanted with human cells or tissues have proved very useful and have played a critical role in translational research, and we think of these mice as one of the real "humanized mice". Furthermore, by application of bioluminescence imaging (BLI) techniques in our humanized mouse, the hormone-dependent behavior of the endometrium regenerated from lentivirally-engineered endometrial cells expressing a variant luciferase can be assessed noninvasively and quantitatively [32]. Our animal model will provide a powerful tool to study the physiology and pathophysiology of human endometrium and also to validate the effect of novel therapeutic agents and gene targeting on endometrium-derived diseases such as endometriosis.

2 Clinical Imaging Systems

2.1 Ultrasound Imaging

Ultrasound imaging, also called ultrasound scanning or sonography, is the most widely used cross-sectional imaging modality in the world [28], and not only medical doctors but also researchers are already trained in the basic principles of ultrasound imaging. Ultrasound images are captured in real-time, and they can show the size, structure, and movement of the internal organs in addition to blood flow. High-frequency sound waves are transmitted to the body, and the returning echoes are recorded to visualize the inside of the body.

Typical diagnostic ultrasound scanners operate in the frequency range of 2-15 MHz. The choice of frequency is a trade-off between spatial resolution and imaging depth. Lower frequencies produce less resolution but are able to image deeper into the body. Ultrasound examinations do not use ionizing radiation (such as X-rays) and avoid pain and tissue damage.

Moreover, ultrasonic visualization of living tissue at microscopic resolution is currently known as ultrasound biomicroscopy (UBM) [12], and UBM is in particular the key imaging tool for embryonic mouse research. UBM transducers operate at 40-100 MHz center frequencies, in contrast to diagnostic clinical transducers. Importantly, imaging issues that arise with high frequencies include loss of penetration, loss of depth of field, and changes in the ultrasound backscatter from blood [9, 12, 13, 42].

2.2 Computerized Tomography

Computerized tomography (CT) is an X-ray technique employing tomography in which digital geometry processing is used to image internal organs of the body. It is relatively safe, painless, and rapid. An X-ray tube, rotating around a specific area of the body, delivers an appropriate amount of X-radiation and takes pictures of that part of the internal anatomy from different angles. CT imaging relies on the principle that various tissue types differentially absorb X-rays as they pass through the body. Modern scanners allow a large series of plane cross-sectional images to be reformatted in various planes or even as three-dimensional representations of structures. Electron beam CT (also called ultrafast CT) [36] is able to take pictures in a tenth of a second. It is useful in creating images of moving parts, such as the heart, without blurring.

CT systems for small animal have been developed specifically for high anatomic resolution imaging [3, 41]. As the relatively low X-ray photon-energy source of 25-50 keV is used, a high-resolution detector system rotates around the animal body to capture images. A typical scan of an entire mouse at a resolution of 100 μm would take about 15 min. Higher spatial resolution requires a longer period of scanning [25].

CT probes (probably iodine- or barium-based for X-ray contrast) have been designed for molecular imaging and used concurrently with CT scanning. Despite the superior soft tissue discrimination of animal scanner, poor soft tissue contrast still necessitates the use of a contrast agent to delineate clearly the internal organs of the animal. However, the use of contrast agent produces an ionization effect that results in radiation damage via superoxides and free radicals. The sensitivity and spatial resolution are dependent on the duration of radiation exposure (scanning time) and the amount of contrast agent used, which respectively affect the body being imaged [25].

2.3 Magnetic Resonance Imaging

Magnetic resonance imaging (MRI) is a diagnostic scanning system that measures the response of the atomic nuclei of body tissues to high-frequency radio waves when the tissues are placed in a strong magnetic field and that produces cross-sectional images of the internal organs. MRI uses a large magnet to generate a magnetic field around the subject. The magnetic field causes hydrogen atoms to align themselves in water and organic compounds, creating what is known as a magnetic dipole. The specific radiofrequency (RF) coils inside the bore of the magnet generate a temporary RF pulse, capable of changing the alignment of these dipoles. Once the pulse ceases, the dipoles return or "relax" to their normal baseline alignment. The relaxation behavior of the dipoles is described by both T1 and T2 relaxation. Both parameters are different for different tissue, resulting in contrast in MRI imaging. Depending on the timing of sequence the contrast can be predominantly T1- or T2 weighted [25].

MRI is primarily used in medical imaging to demonstrate pathological or other physiological alterations of living tissues. MRI can extract not only structural information but also physiological and molecular information. These are helpful in the diagnosis of abnormalities without the possibly harmful effects of X-rays or other forms of radiation. MRI scans are very valuable in detecting and delineating tumors and in providing images of the brain, the spinal cord, the heart, and other soft-tissue organs. The disadvantage is that it requires a longer scanning time than other computer-assisted forms of scanning, which makes it more sensitive to motion and of less value in scanning the chest or abdomen. Although the images are similar to those of CT scans, MRI images provide better contrast between normal and diseased tissue than those produced by other computer-assisted imaging.

When it is used to provide a dynamic picture of oxygen metabolism during specific mental activities, it is called functional MRI (fMRI) [11]. This shows changes of local blood flow and hemoglobin oxygenation in response to altered neuronal activity. This change correlates with levels of neuronal activity in specific brain regions, and therefore fMRI allows mapping of functional centers of the brain.

High-resolution MRI is widely used in small animals (micro-MRI, μMRI), and reveals fine morphological details [37]. Stronger magnetic fields can be used and higher spatial resolution (25-50 μm) with exquisite morphological detail can be achieved, but for the most part requires field strengths of 7-11 T and long acquisition times (hours-typically overnight). Advantages of μMRI include its noninvasive nature (low toxicity), excellent tissue contrast, and ability to reconstruct images in any plane, including 3D reconstruction [42].

Recently, developments in animal MRI have focused on the development of new contrast agents that increase sensitivity and specificity. Contrast agents can be classified as nonspecific, targeted, and smart probes [4]. Nonspecific probes such as gadolinium chelates show a nonspecifically distributed pattern and are used to measure tissue perfusion and vascular permeability. Targeted probes such as gadolinium-labeled avidin and annexin V-supramagnetic iron oxide nanoparticles are designed to specifically bind to ligands such as peptides and antibodies. Smart probes tag a specific ligand similar to targeted agents but differ in that the probe signal changes on interaction with the specific ligand [25].

2.4 Single-Photon Emission Computed Tomography

Single-photon emission computed tomography (SPECT) is a nuclear medicine tomographic imaging technique using gamma rays (γ-rays). It allows us to visualize blood flow and metabolism. A radioactive isotope is attached to a substance that is easily taken up by target cells. As the isotope breaks down, it releases energy in the form of γ-rays. The γ-rays are like beacons of light that signal where the compound is in the body and are acquired by a gamma camera from multiple angles. A computer then translates these data to yield a 3D data set. This data set can be freely reformatted or manipulated to show cross-sectional slices along any chosen axis of the body.

In contrast to clinical use, small animal imaging require higher spatial resolution. This is achieved by pinhole collimators, and SPECT systems for small animals have appeared in recent years [2, 22].

SPECT is similar to a positron emission tomography (PET) scan at first glance, but it differs from PET scans in that isotopes are direct gamma emitters in a single direction, necessitating different instrumentation for detection, which results in a limitation to the detection efficiency (to around 10^{-4} of number of γ-rays) of SPECT. Longer-lived radioactive isotopes, including ^{111}In, ^{123}I, ^{125}I, ^{201}Tl, and ^{99m}Tc, are typically used [27, 45].

2.5 Positron Emission Tomography

PET is a clinical imaging technique that monitors metabolic, or biochemical, activity in the brain and other organs by tracking the movement and concentration of a radioactive tracer injected into the bloodstream. The radioactive atoms used in a PET scan emit subatomic particles called positrons (positive electrons), which collide with their negatively charged counterparts, namely, electrons. The two particles annihilate each other and emit two 511-keV photons (γ-rays) that radiate in opposite directions and can be recorded by a ring of detectors round the body and traced back to their point of origin. The acquired data are organized by a computer into 3D data sets to produce two-dimensional slices for all angular views.

In the past, PET was only used for large animals. However, because of technological innovation and improved imaging resolution in recent years, micro-PET scanners have been developed for small-animal imaging [16, 20]. The major limitations of PET are its spatial resolution and image noise. Spatial resolution of PET scans is typically about 2^3 mm^3 [8]. Newer-generation scanners can achieve a resolution of about 1^3 mm^3 [6].

There are many PET isotopes with different half-times ($T_{1/2}$) from minutes to days, for example, ^{15}O, ^{13}N, ^{11}C, ^{18}F, ^{64}Cu, and ^{124}I [27, 45]. A well-known example of an isotope, 18-fluorodeoxyglucose (^{18}FDG), is widely applied in tumor studies. It accumulates in tumor-specific sites because tumor cells have greater glucose uptake rate and glycolytic metabolism than normal tissues [17]. Many other radiopharmaceuticals can be engineered, most commonly from "biologic" positron emitters such as ^{18}F and ^{11}C, to target specific molecular targets within defined in vivo biochemical pathways and processes [7].

3 Optical Imaging Systems

Optical imaging systems have been developed that use both bioluminescent [44] and fluorescent [15] signals. This technique employs quantitative light emission, namely, photons, to obtain measurements of relevant biological parameters, including proteins and nucleic acids. Further advancement has come through the development of new targeted bioluminescent probes, near-infrared fluorochromes, and red fluorescent proteins [52]. To detect low levels of light or photons, a very sensitive charge-coupled device (CCD) detector is used [21]. The CCD detector is silicon-based and is capable of detecting light from the visible range (395- 600 nm) to the near-infrared range (NIR: 600-2,500 nm) [25]. This imaging modality shows attractive data from organs close to skin in small animals, but this is likely to remain restricted to relatively superficial targets because of the absorbing and scattering properties of tissue in the visible range and NIR [26].

3.1 Fluorescence Imaging

Fluorescence is light of a visible color emitted from a substance under stimulation or excitation by light or other forms of electromagnetic radiation. The light is given off only while the stimulation continues. Visible light is used to excite fluorescence within the subject, and a camera or fluorescence microscopy system detects the emitted light from the region of interest. The commonly used strategy is to fluorescently tag the cells, tissue, or molecules under investigation with substances known to fluoresce. The most popular fluorochrome is green fluorescent protein (GFP), which is derived from the jellyfish *Aequorea victoria*. The wild-type GFP emits light at 509 nm, whereas its variant EGFP has a longer emitting wavelength and is 35-fold brighter [24]. Thus there are a lot of fluorescent proteins, and increasing numbers of new bright fluorescent probes with a variety range of emission wavelengths and greater stability have been developed and are now available [46].

Fluorochromes of wavelengths greater than 600 nm should be used in order to minimize absorbance by surrounding tissue and to distinguish background and autofluorescence [49]. Indeed, the use of NIR fluorochromes achieves maximum tissue penetration and minimum background and autofluorescence [53], and several applications have exploited the NIR range [29] with best results typically achieved when the emission wavelengths of the dye are between 500 and 950 nm [30].

3.2 Bioluminescence Imaging

Bioluminescence is biochemical emission of light, with very little heat, by living organisms such as fireflies and deep-sea fishes as the result of a chemical reaction during which chemical energy is converted to light energy. This reaction is mediated by one of the luciferase family of photoproteins that can be isolated either from the sea pansy (*Renilla reniformis*) or from the North American firefly (*Photimus pyralis*). Different organisms produce different bioluminescent substances and use different substrates. Marine bioluminescent organisms use coelenterazine as a substrate, and terrestrial organisms use d-luciferin, which provides a longer-lived and longer wavelength. Luciferase is normally bound to ATP (adenosine triphosphate) in an inactive form, but in catalysis of luciferin it is liberated from the ATP and combines with oxygen to form an oxyluciferin in an excited state, which quickly decays, emitting photons of visible light as it does [54].

As in the case of fluorescence imaging (FLI), the use of luciferases that have a significant portion of their emission greater than 600 nm, such as luciferase derived from fireflies and click beetles (approximately 60% of the light emitted from these two enzymes has wavelengths greater than 600 nm), will lead to more sensitive detection of the labeled cells in vivo [38, 55].

In vivo bioluminescence imaging (BLI) has been applied in the assessment of the extent of tumor growth and response to therapy by transplantation of tumor cells transfected to express luciferase into animals [47]. Furthermore, the expression of luciferase can be controlled so that it is only expressed when a gene of interest is being transcribed [5, 31]. Cells expressing luciferase in animal models can be easily identified through their emission of light in the range of 400-620 nm by administration of luciferin [19]. This technology has become an invaluable tool that has been employed to dynamically monitor tumor growth or transcriptional activity in living animals.

3.3 Comparison of BLI with FLI

First of all, as the luciferase reaction is energy-dependent and requires ATP and oxygen, the luminescent signal is produced only from living cells, which are different from the fluorescent signals that can be sustained even in nonliving cells. The major attraction of BLI over FLI is that, although absolute light levels generated by the targets may be low, photons are generated generally only where luciferase is present, leading to an extremely low level of background signals and excellent signal-to-noise ratios (SNRs). In contrast, FLI requires an external light source to stimulate the emission of light from the probe, and the light source could also generate bright background signals arising from the animal's intrinsic autofluorescence [26, 44, 50].

However, there are some limitations in BLI. First, light transmission efficiency is dependent on the type and location of tissue being assessed because of the narrow range, from 400 nm to 620 nm, of the light emission peak. Highly vascular organ structures contain hemoglobin that absorbs transmitted light, which results in about a ten-fold reduction of the bioluminescent signal for every centimeter of tissue depth [10]. Second, because the catalytic reaction in BLI is time- and enzyme-dependent, the window period for optimum image capture must be determined [25].

In contrast, a distinctive advantage of FLI over BLI is that it does not require administration of a substrate for visualization. FLI may be more convenient and allow easier capture of images at multiple time points, since administration of a substrate into the animal is not required. In addition, FLI is a more flexible technology, since it permits the use of a far wider range of probes, labeling methods, and targets. The number of photons emitted in FLI is orders of magnitude greater than that in BLI [26, 50].

Among the great number of fluorescent probes, Quantum Dots (Q-Dots), which are semiconductor nanocrystals that have long-term stability and fluoresce brightly up to the NIR spectrum on excitation, have been developed for imaging [1]. Q-dots enable antibody-targeted spectral imaging and analysis such as examining the distribution of Q-dot-labeled antitumor antibodies in mice [14]. In addition, this method will be available as multimodal contrast agents for not only FLI but also PET or MRI detection [33]. An agent such as Q-Dots is valuable in multimodality imaging as described below.

4 Multimodality Imaging

Imaging modalities can be divided into two groups. One group includes ultrasound, CT, and MRI, providing structural information, while another group includes SPECT, PET, and optical imaging, providing functional or molecular information. Both groups have some drawbacks and advantages (Tables 1 and 2). Therefore, the combination of different imaging modalities has been developed to offset the disadvantage of each modality, and it will develop into a powerful tool with the recent development of contrast agents including Q-Dots as mentioned in the previous section.

PET/CT has only recently developed to enable accurate diagnosis. By combining the structural anatomic information of CT scans with the metabolic or cellular activity data of PET scans, it has become possible to visualize anatomy and function simultaneously. The anatomic information enables compensation of the correlated radionuclide data for physical perturbations such as photon attenuation, scatter radiation, and partial volume errors. Thus, dual-modality imaging provides a priori information that can improve both the visual quality and the quantitative accuracy of the radionuclide images. The hybrid imaging of PET/CT has been shown to improve not only the sensitivity of PET interpretation but also its specificity. Micro-PET/CT is used for small animals to obtain high anatomic resolution with functional information [48, 49]. The combination of the various modalities, called multimodality imaging, offers valuable information. By the same token, SPECT/CT [18] and optical PET [43] have been developed in addition to PET/CT.

A further development in FLI is fluorescence molecular tomography (FMT), which has been developed for acquiring images of fluorescently labeled proteins and for deeper targets. It employs continuous wave or pulsed light from different sources to excite the fluorochrome label, and multiple detectors are arranged spatially around the subject analogous to the set-up in CT or MR scanners [39, 40]. Computation of the data generates a 3D image. The resulting images have a resolution of 1-2 mm, and the fluorochrome detection threshold is in the nanomolar range [25].

5 A Novel In Vivo Imaging System of the Endometrial Model Mouse

As the bioluminescent signal is emitted only from living cells expressing luciferase, leading to low background noise, we chose BLI as the modality to visualize an artificial menstrual cycle in mice and developed an in vivo imaging system based on this concept.

Table 1 The properties of currently available in vivo imaging techniques

Modality	Basis	Reagents	Acquisition time	Tissue penetration depth	Spatial resolution	Signal quantification capabilities	Cost (equipment and usage)
Ultrasound	High-frequency sound waves	Microbubbles	minutes	1–200 mm	50–500 μm (animal) 0.1–1 mm (clinical)	Low	Low
CT	X-rays	Iodine	minutes	No limit	30–50 μm (animal) 0.5–1 mm (clinical)	N/A	Medium-high
MRI	Radio frequency waves	Paramagnetic cation probes	minutes–hours	No limit	25–100 μm (animal) 0.2 mm (clinical)	Medium	High
PET	High energy γ-rays	^{18}F, ^{11}C, ^{13}N, ^{15}O, ^{124}I, ^{94m}Tc labeled probes or substrates for reporter transgenes	minutes	No limit	1–2 mm (animal) 6–10 mm (clinical)	High	High
SPECT	Low energy γ-rays	^{99m}Tc, ^{111}In, ^{125}I labeled probes	minutes	No limit	1–2 mm (animal) 7–15 mm (clinical)	Medium-high	Medium-high
FLI	Visible to near-infrared light	Fluorescent proteins, fluorescent dyes, and quantum dots (semiconductor)	seconds–minutes	1–20 mm	1–10 mm (animal) (depending on tissue depth)	Low-medium	Low
BLI	Visible light	Luciferase and substrates (luciferin, coelenterazine)	seconds–minutes	1–10 mm	1–10 mm (animal) (depending on tissue depth)	Low-medium	Low

In Vivo Imaging in Humanized Mice 189

Table 2 Advantages and disadvantages of in vivo imaging modalities

Modality	Advantages	Disadvantages
Ultrasound	Real-time imaging, low cost, and user-friendly	Limited ability to image through bone or lungs
CT	Good anatomic resolution	Relatively poor soft-tissue contrast
		Radiation to animal with CT contrast agents
MRI	Highest spatial resolution	Low sensitivity
	The ability to combine functional information and anatomic details	Long acquisition time
		Long image processing time
PET	High sensitivity	Low resolution
	The ability of quantitative measure	Unincorporated substrate can increase noise
	Variety of probes and strategies confers a high degree of versatility	Cyclotron required to generate short-lived radioisotope
		Radiation to animal
SPECT	Multiple probes can be detected simultaneously	10–100x less sensitive than PET
		Relative low resolution
	Radioisotopes have longer half-lives than those used in PET	Radiation to animal
FLI	High sensitivity	Prone to attenuation with increased tissue depth
	Easy and quick to image	
	Multiple reporter wavelengths enables multiplex imaging	Probes with emission wavelength <600 nm prone to autofluorescence of nonlabeled cells
	Detect fluorochromes in live and dead cells	
	Transgene-based approach confers versatility	
BLI	High sensitivity	Low anatomic resolution
	Easy and quick to image	Light emission prone to attenuation with increased tissue depth
	Provides relative measure of cell viability or function	
	Available for gene expression and cell tracking	
	Transgene-based approach confers versatility	

5.1 Lentiviral Introduction of Reporter Genes into Primary Endometrial Cells

We have developed a recombinant lentivirus capable of introducing and stably expressing both the Venus [a yellow fluorescent protein (YFP) mutant] [35] gene and the click beetle red-emitting luciferase (CBR luc, a luciferase variant) [55] gene in the targeted cells (Fig. 1a). These two reporter markers are useful for

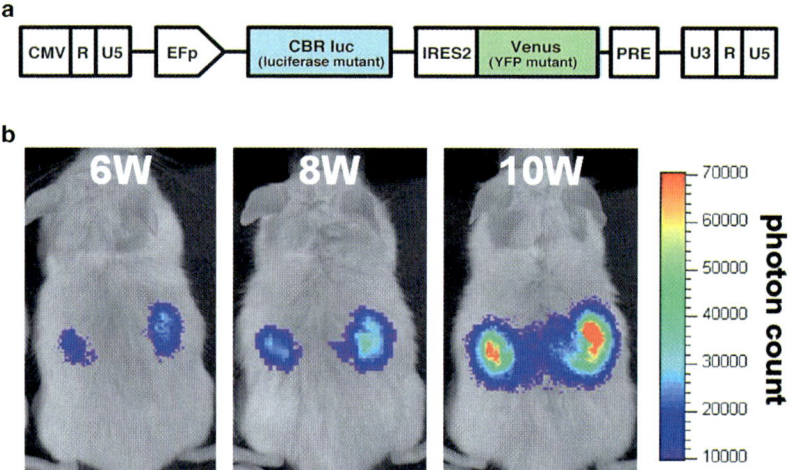

Fig. 1 Lentiviral construct for fluorescence and bioluminescence and optical bioluminescence images of the growth of the tissues reconstructed from lentivirally transduced SDECs in living NOG mice. **a** Lentiviral construct encoding a dual function CBR luc (a luciferase variant) and Venus (a YFP variant) bicistronic reporter gene connected via an internal ribosomal entry site (*IRES*). **b** Representative sequential BLI of NOG mice treated for different durations with two E_2 pellets

flow cytometry sorting and the detection of living cells from outside the body, respectively. Indeed, fluorescence microscopy-revealed Venus signals were detected in lentivirally-infected cells, and these cells could be sorted easily by flow cytometry. Moreover, CBR luc (maximum emission at 613 nm) has the potential to pass through thicker tissue. By combining the advantages of the lentivirus and CBR luc, we successfully assessed the dynamic state of the endometrial reconstructs in living NOD/SCID/γ_c^{null} (NOG) mice [23].

5.2 Noninvasive, Real-Time, and Quantitative Assessment of the Reconstructs by BLI

We transplanted singly dispersed endometrial cells (SDECs) beneath the kidney capsules of severely immunodeficient mice, NOG mice, and we demonstrated for the first time that a functional endometrium-like structure can be regenerated from SDECs [32]. This model is the humanized mouse we developed. To apply this humanized mouse to an in vivo BLI system, we transplanted the human endometrial cells infected with our above-mentioned lentivirus beneath the kidney capsule. Consequently, sequential BLI of the ventrally positioned estradiol (E_2)-treated NOG mouse 6-10 weeks after xenotransplantation revealed bioluminescent (CBR) signals in locations corresponding to the bilateral kidneys (Fig. 1b).

Furthermore, we showed that the signal intensities reflecting the volume of the reconstructed tissue were enhanced in an E_2 dose- and time-dependent manner by using ovariectomized NOG (OVX-NOG) mice without or with one or two long-term continuous release pellets of E_2 (Fig. 2a). This means that in our system the growth behavior of the reconstructed tissue could be assessed quantitatively and sequentially.

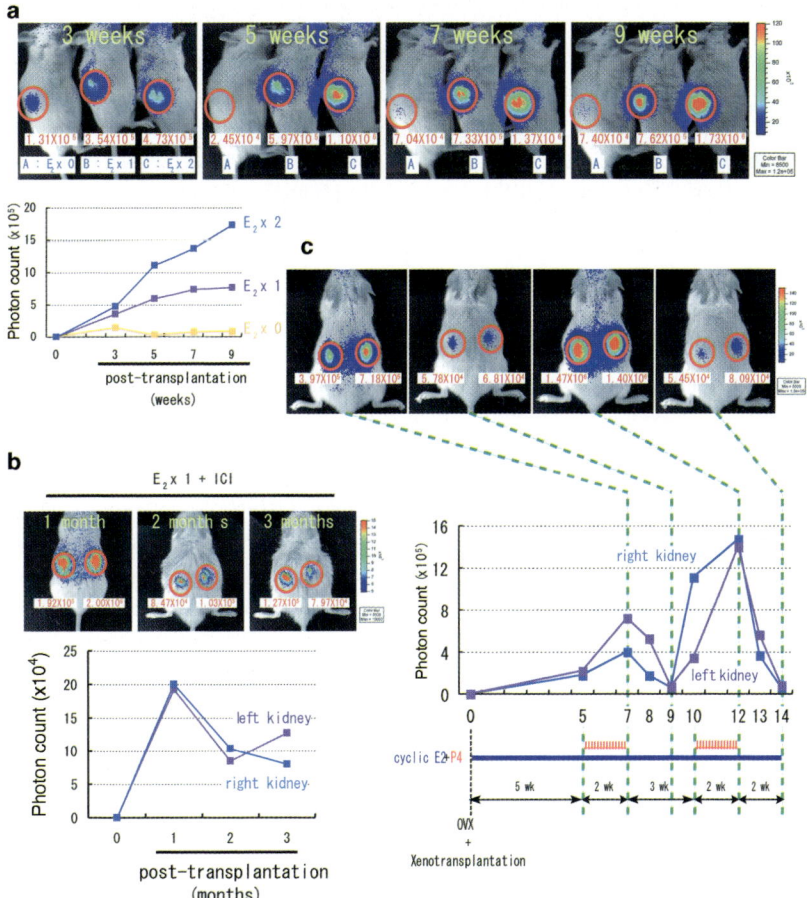

Fig. 2 Quantitative assessment of the growth of the regenerated endometrium. **a** Representative BLI (*top panels*) and serial photon count measurements (*bottom panel*) of NOG mice treated for different durations with the various indicated doses of E_2 pellets. The photon count value of each region of interest (ROI, *red circle*) is indicated. **b** Representative BLI (*top panels*) and serial photon count measurements (*bottom panel*) of a xenotransplanted OVX-NOG mouse treated with E_2 in combination with daily injections of ICI 182,780, a pure estrogen antagonist. **c** Representative sequential BLI (*top panels*) and serial photon count measurements (*middle panel*) of a xenotransplanted OVX-NOG mouse undergoing cyclic E_2+P_4 treatment (*bottom panel*) to induce artificial menstrual cycle-related changes. Images are adapted, with permission, from Ref. 53 Copyright 2007, National Academy of Sciences, USA

Additionally, in contrast to the xenotransplanted mice treated with E_2 alone, the signal intensity was not increased but rather decreased 2-3 months after cotreatment with E_2 in combination with ICI-182,780 (ICI), a pure estrogen antagonist [51] (Fig. 2b). These data indicated that the antagonistic effect of ICI can be noninvasively and successfully assessed and this system could be used as a tool for drug screening.

Finally, we monitored the dynamic changes of the endometrial reconstructs during an artificial menstrual cycle induced by cyclic treatment with E_2 in combination with P_4 ($E_2 + P_4$ treatment). Sequential BLI revealed that the signal intensities fluctuated dramatically in accord with the addition and withdrawal of progesterone (P_4) (Fig. 2c). In particular, tissue breakdown and regression after P_4-withdrawal and subsequent tissue regeneration faithfully reflected the decrease and increase in signal intensity, respectively (Fig. 2c). Thus, the repeated menstrual cycle-related changes of the transplants can be successfully reproduced and also noninvasively monitored in living NOG mice.

5.3 Discussion of This System

This animal model has several advantages over the current models of the endometrium and endometriosis.

First, the transplants are singly dispersed cells isolated from the human endometrium. A single-cell suspension is adequate for experimental procedures such as cell selection, genetic engineering, and quantitative assessment as compared to dissected sections. Quantification of the transplanted cells and the homogeneity of each model animal are especially and critically important for the comparative analysis of therapeutic agents. One potential disadvantage of singly dissociated cells is unpredicted scattering and spreading of the cells transplanted subcutaneously or intraperitoneally, making it difficult to identify the transplanted site and to evaluate the reconstructs. However, this limitation has been overcome by transplantation beneath the kidney capsule. Kidney capsule transplantation has the further advantage of being ideal for BLI assessment, not only of endometrial cells but other cell types as well, for example, (1) evaluation of cell type-dependent responses to tested drugs when two different types of cells are separately transplanted into each kidney, (2) easy macroscopic identification of the small reconstruct, and (3) efficient detection of the bioluminescent signals derived from the transplant because of the relatively superficial location on the kidney.

Second, the regenerated tissues in our model exhibit abundant vascularization, endometrial cell components, and tissue organization, all of which result in the long-term maintenance of the reconstructed structure and the hormone-dependent changes characteristic of human cycling endometrium and/or endometriotic explants. It is suitable for the evaluation of the effects of therapeutic reagents.

Third, the transplants can be assessed for a prolonged period in a noninvasive, real-time, and quantitative manner. Our lentiviral vector system [34] enables primary endometrial cells stably and permanently to express CBR luc, the light emitted from which has the capacity to pass through thicker tissues. Because of the advantages of the lentivirus and CBR luc, we were able to assess successfully the dynamic state of the endometrial reconstructs in living NOG mice.

By combining the unique potential of the human endometrium together with the special characteristics of NOG mice and lentivirus-mediated cell engineering, we are able to report on the first animal model suitable for the study of endometrial physiology/pathophysiology and the pathogenesis of endometriosis through non-invasive, real-time, and quantitative assessment of ectopically reconstituted endometrium-like tissues. Furthermore, this animal model system, based on the strategy of transplanting lentivirally engineered cells expressing a bioluminescent marker beneath the kidney capsule, can be potentially applicable for drug testing and gene target validation not only in endometriosis, but also in other various types of neoplastic disease.

Finally, BLI will provide novel methods to analyze biological processes, and it has a huge potential. In the near future, it will surely evolve and be developed to enable the visualization of two or three populations of cells simultaneously and to quantify cell numbers accurately depending on the innovative properties of detection systems and bioluminescent compounds. The visualization of many populations and improved quantification allow more complex kinetic analysis. Imaging studies using BLI in humans are limited, but information from the use of BLI in humanized mice will have a potentially great effect on clinical medicine. BLI will contribute to the modification and evaluation of preclinical trials, especially in the field of cell-based therapeutics, which will almost certainly demonstrate exponential expansion in the near future.

6 Conclusion

Recent animal models of human diseases (i.e., humanized mice) and in vivo imaging technologies are improving. These technologies will provide the opportunity for a new aspect in the field of animal experiments, delivering novel information and important insights.

Noninvasive and real time assessment with high sensitivity, accurate quantification, and high spatial resolution is ideal for imaging. In addition, the ideal modality requires obtaining images simply and in a short capture time.

Unfortunately, no modality and no animal model developed to date could fulfill all these requirements at once. However, the technology of each modality is improving every second. The combination of different imaging modalities (e.g., SPECT/CT, PET/CT, and FMT) is a novel powerful tool. Therefore, it is important to be familiar with in vivo imaging methods in humanized mice and to be able to choose suitable modalities as the occasion demands.

References

1. Balaban RS, Hampshire VA (2001) Challenges in small animal noninvasive imaging. ILAR J 42:248-62
2. Beekman FJ, van der Have F, Vastenhouw B, van der Linden AJ, van Rijk PP, Burbach JP, Smidt MP (2005) U-SPECT-I: a novel system for submillimeter-resolution tomography with radiolabeled molecules in mice. J Nucl Med 46:1194-200
3. Berger F, Lee YP, Loening AM, Chatziioannou A, Freedland SJ, Leahy R, Lieberman JR, Belldegrun AS, Sawyers CL, Gambhir SS (2002) Whole-body skeletal imaging in mice utilizing microPET: optimization of reproducibility and applications in animal models of bone disease. Eur J Nucl Med Mol Imaging 29:1225-36
4. Bremer C, Ntziachristos V, Weissleder R (2003) Optical-based molecular imaging: contrast agents and potential medical applications. Eur Radiol 13:231-43
5. Cao YA, Wagers AJ, Beilhack A, Dusich J, Bachmann MH, Negrin RS, Weissman IL, Contag CH (2004) Shifting foci of hematopoiesis during reconstitution from single stem cells. Proc Natl Acad Sci USA 101:221-6
6. Chatziioannou A, Tai YC, Doshi N, Cherry SR (2001) Detector development for microPET II: a 1 microl resolution PET scanner for small animal imaging. Phys Med Biol 46:2899-910
7. Chatziioannou AF (2002) Molecular imaging of small animals with dedicated PET tomographs. Eur J Nucl Med Mol Imaging 29:98-114
8. Cherry SR, Gambhir SS (2001) Use of positron emission tomography in animal research. ILAR J 42:219-32
9. Coatney RW (2001) Ultrasound imaging: principles and applications in rodent research. ILAR J 42:233-47
10. Contag CH, Contag PR, Mullins JI, Spilman SD, Stevenson DK, Benaron DA (1995) Photonic detection of bacterial pathogens in living hosts. Mol Microbiol 18:593-603
11. DeYoe EA, Bandettini P, Neitz J, Miller D, Winans P (1994) Functional magnetic resonance imaging (FMRI) of the human brain. J Neurosci Methods 54:171-87
12. Foster FS, Pavlin CJ, Harasiewicz KA, Christopher DA, Turnbull DH (2000) Advances in ultrasound biomicroscopy. Ultrasound Med Biol 26:1-27
13. Foster FS, Zhang MY, Zhou YQ, Liu G, Mehi J, Cherin E, Harasiewicz KA, Starkoski BG, Zan L, Knapik DA, Adamson SL (2002) A new ultrasound instrument for in vivo microimaging of mice. Ultrasound Med Biol 28:1165-72
14. Gao X, Cui Y, Levenson RM, Chung LW, Nie S (2004) In vivo cancer targeting and imaging with semiconductor quantum dots. Nat Biotechnol 22:969-76
15. Graves EE, Ripoll J, Weissleder R, Ntziachristos V (2003) A submillimeter resolution fluorescence molecular imaging system for small animal imaging. Med Phys 30:901-11
16. Green MV, Seidel J, Vaquero JJ, Jagoda E, Lee I, Eckelman WC (2001) High resolution PET, SPECT and projection imaging in small animals. Comput Med Imaging Graph 25:79-86
17. Griffin JL, Shockcor JP (2004) Metabolic profiles of cancer cells. Nat Rev Cancer 4:551-61
18. Hasegawa BH, Iwata K, Wong KH, Wu MC, Da Silva AJ, Tang HR, Barber WC, Hwang AH, Sakdinawat AE (2002) Dual-modality imaging of function and physiology. Acad Radiol 9:1305-21
19. Hastings JW (1996) Chemistries and colors of bioluminescent reactions: a review. Gene 173:5-11
20. Herschman HR (2003) Micro-PET imaging and small animal models of disease. Curr Opin Immunol 15:378-84
21. Honigman A, Zeira E, Ohana P, Abramovitz R, Tavor E, Bar I, Zilberman Y, Rabinovsky R, Gazit D, Joseph A, Panet A, Shai E, Palmon A, Laster M, Galun E (2001) Imaging transgene expression in live animals. Mol Ther 4:239-49

22. Ishizu K, Mukai T, Yonekura Y, Pagani M, Fujita T, Magata Y, Nishizawa S, Tamaki N, Shibasaki H, Konishi J (1995) Ultra-high resolution SPECT system using four pinhole collimators for small animal studies. J Nucl Med 36:2282-7
23. Ito M, Hiramatsu H, Kobayashi K, Suzue K, Kawahata M, Hioki K, Ueyama Y, Koyanagi Y, Sugamura K, Tsuji K, Heike T, Nakahata T (2002) NOD/SCID/γ_c^{null} mouse: an excellent recipient mouse model for engraftment of human cells. Blood 100:3175-82
24. Kaneko K, Yano M, Yamano T, Tsujinaka T, Miki H, Akiyama Y, Taniguchi M, Fujiwara Y, Doki Y, Inoue M, Shiozaki H, Kaneda Y, Monden M (2001) Detection of peritoneal micrometastases of gastric carcinoma with green fluorescent protein and carcinoembryonic antigen promoter. Cancer Res 61:5570-4
25. Koo V, Hamilton PW, Williamson K (2006) Non-invasive in vivo imaging in small animal research. Cell Oncol 28:127-39
26. Levenson RM, Mansfield JR (2006) Multispectral imaging in biology and medicine: slices of life. Cytometry A69:748-58
27. Levin CS (2005) Primer on molecular imaging technology. Eur J Nucl Med Mol Imaging 32 Suppl 2: S325-45
28. Liang HD, Blomley MJ (2003) The role of ultrasound in molecular imaging. Br J Radiol 76 Spec No 2: S140-50
29. Lin Y, Weissleder R, Tung CH (2002) Novel near-infrared cyanine fluorochromes: synthesis, properties, and bioconjugation. Bioconjug Chem 13:605-10
30. Mansfield JR, Gossage KW, Hoyt CC, Levenson RM (2005) Autofluorescence removal, multiplexing, and automated analysis methods for in-vivo fluorescence imaging. J Biomed Opt 10:41207
31. Massoud TF, Paulmurugan R, Gambhir SS (2004) Molecular imaging of homodimeric protein-protein interactions in living subjects. FASEB J 18:1105-7
32. Masuda H, Maruyama T, Hiratsu E, Yamane J, Iwanami A, Nagashima T, Ono M, Miyoshi H, Okano HJ, Ito M, Tamaoki N, Nomura T, Okano H, Matsuzaki Y, Yoshimura Y (2007) Noninvasive and real-time assessment of reconstructed functional human endometrium in NOD/SCID/gamma c^{null} immunodeficient mice. Proc Natl Acad Sci USA 104:1925-30
33. Michalet X, Pinaud FF, Bentolila LA, Tsay JM, Doose S, Li JJ, Sundaresan G, Wu AM, Gambhir SS, Weiss S (2005) Quantum dots for live cells, in vivo imaging, and diagnostics. Science 307:538-44
34. Miyoshi H, Blomer U, Takahashi M, Gage FH, Verma IM (1998) Development of a self-inactivating lentivirus vector. J Virol 72:8150-7
35. Nagai T, Ibata K, Park ES, Kubota M, Mikoshiba K, Miyawaki A (2002) A variant of yellow fluorescent protein with fast and efficient maturation for cell-biological applications. Nat Biotechnol 20:87-90
36. Nasir K, Budoff MJ, Post WS, Fishman EK, Mahesh M, Lima JA, Blumenthal RS (2003) Electron beam CT versus helical CT scans for assessing coronary calcification: current utility and future directions. Am Heart J 146:969-77
37. Natt O, Watanabe T, Boretius S, Radulovic J, Frahm J, Michaelis T (2002) High-resolution 3D MRI of mouse brain reveals small cerebral structures in vivo. J Neurosci Methods 120:203-9
38. Negrin RS, Contag CH (2006) In vivo imaging using bioluminescence: a tool for probing graft-versus-host disease. Nat Rev Immunol 6:484-90
39. Ntziachristos V, Tung CH, Bremer C, Weissleder R (2002) Fluorescence molecular tomography resolves protease activity in vivo. Nat Med 8:757-60
40. Ntziachristos V, Weissleder R (2002) Charge-coupled-device based scanner for tomography of fluorescent near-infrared probes in turbid media. Med Phys 29:803-9
41. Paulus MJ, Gleason SS, Easterly ME, Foltz CJ (2001) A review of high-resolution X-ray computed tomography and other imaging modalities for small animal research. Lab Anim (NY) 30:36-45
42. Phoon CK (2006) Imaging tools for the developmental biologist: ultrasound biomicroscopy of mouse embryonic development. Pediatr Res 60:14-21

43. Prout DL, Silverman RW, Chatziioannou A (2004) Detector concept for OPET-A combined PET and optical imaging system. IEEE Trans Nucl Sci 51:752-756
44. Rice BW, Cable MD, Nelson MB (2001) In vivo imaging of light-emitting probes. J Biomed Opt 6:432-40
45. Shah K, Jacobs A, Breakefield XO, Weissleder R (2004) Molecular imaging of gene therapy for cancer. Gene Ther 11:1175-87
46. Shaner NC, Steinbach PA, Tsien RY (2005) A guide to choosing fluorescent proteins. Nat Methods 2:905-9
47. Sweeney TJ, Mailander V, Tucker AA, Olomu AB, Zhang W, Cao Y, Negrin RS, Contag CH (1999) Visualizing the kinetics of tumor-cell clearance in living animals. Proc Natl Acad Sci USA 96:12044-9
48. Townsend DW (2001) A combined PET/CT scanner: the choices. J Nucl Med 42:533-4
49. Townsend DW, Cherry SR (2001) Combining anatomy and function: the path to true image fusion. Eur Radiol 11:1968-74
50. Troy T, Jekic-McMullen D, Sambucetti L, Rice B (2004) Quantitative comparison of the sensitivity of detection of fluorescent and bioluminescent reporters in animal models. Mol Imaging 3:9-23
51. Wakeling AE, Dukes M, Bowler J (1991) A potent specific pure antiestrogen with clinical potential. Cancer Res 51: 3867-73
52. Weissleder R (2001) A clearer vision for in vivo imaging. Nat Biotechnol 19:316-7
53. Weissleder R (2002) Scaling down imaging: molecular mapping of cancer in mice. Nat Rev Cancer 2:11-8
54. Wilson T, Hastings JW (1998) Bioluminescence. Annu Rev Cell Dev Biol 14:197-230
55. Zhao H, Doyle TC, Coquoz O, Kalish F, Rice BW, Contag CH (2005) Emission spectra of bioluminescent reporters and interaction with mammalian tissue determine the sensitivity of detection in vivo. J Biomed Opt 10:41210

Index

A
Acute myeloid leukemia, 66
Aging, 81
Animal model, 180–181, 186, 193–194
Anti-asialo GM1, 56, 171
Antigen presenting cell (APC), 101–102
Anti-IL-2Rb, 56
Asialo GM1, 68
ATL, 141–143
Autoimmunity, 37
AZT, 135

B
B2 cell, 96, 99
b2m^{null} mice, 56
Balb/c-109, 117–118, 120
Beige mice, 54
Bioluminescence imaging (BLI), 179–181, 185–194
BLT, 149, 153–163
Bone marrow, 78

C
Cancer, 38
CB-17-scid mice, 55
CCR5, 138–139
CD5+B1 cell, 96, 99
Cell lines
    melanoma 174
    pancreatic cancer 170–171
CFP, 91
Class switch, 99
Clinical imaging systems, 181
Computerized tomography (CT), 179–183, 187–190
    ultrafast CT, 182
Cytokines, 31

D
Different human hematopoietic lineages, 150–151, 154, 156, 162
DiGeorge/Nezelof syndromes, 149–150, 161
2,4-Dinitrophenylated keyhole limpet hemocyanin (DNP-KLH), 98, 100
Diphtheria toxin, 68
DNA
    microarray 172
    transfection 173
DNA-dependent protein kinase (Prkdc), 55
Double negative (DN), 99
Double positive (DP), 99
Dramatype, 4

E
E-cadherin, 174
Endometriosis, 180–181, 193–194
Endometrium (endometrial), 66, 180–181, 187, 190–194
Engraftment, 36–37
Erythropoietin, 67
Exhaustion, 80

F
FISH, 90–92
Fluorescence-activated cell sorting (FACS), 97
Fluorescence imaging (FLI), 179–180, 185–190
Fluorescence molecular tomography (FMT), 180, 187, 194
Follicular dendritic cells (FDC), 64

G
Gene marking, 79
Genetic background, 9
GFP, 91
GM-CSF, 67
Graft-versus-host-disease (GVHD), 97, 136–137
Granzyme B null mice, 57
Growth factors, 30

H
Hematopoietic microenvironment, 81
Hematopoietic stem cells (HSCs), 78
Hepatitis C virus, 66
Herpes simplex virus thymidine kinase, 68
Historical perspective, 27–28
HIV, 8, 133–134, 138, 143
   infection, 125
     specific immune responses, 128
HLA, 68
Hodkin lymphoma, 66
Hormones, 30
HTLV, 133–134, 141–144
Human DC, 159–160, 162
Human Hemato-Lymphoid System Mice, 125
Human hematopoeisis, 151, 163
Hu-PBL-SCID, 134, 136–137, 140
hu-PBL-SCID mouse model, 66

I
IFN-γ, 59
IgG antibody
   antigen specific, 98, 102
IgM antibody
   antigen specific, 98, 102
IL-15, 111–112, 114–115, 119
IL2 receptor gamma chain, 35
IL2rγ^{null} mice, 57
IL-3, 67
IL-4, 68
IL-5, 68
IL-7, 111–114, 119
Immuno deficient mice, 95, 97
Infectious disease, 37
Innate immunity, 29–30, 41
Intestinal flora, 17
Intra-bone marrow transplantation, 81

K
KIR, 110, 113, 115–116, 119

L
Leakiness, 60
Lentivirus, 190–191, 194
Linear amplification-mediated-PCR, 79
Liver metastasis, 66

M
Magnetic cell sorting (MACS), 97
Magnetic resonance imaging (MRI), 179–180, 182–183, 187–190
   functional-MRI, 180, 183
   micro-MRI 180, 183
Major histocompatibility complex (MHC), 110, 115, 118–120
   classII KO mice, 101
   restriction, 102
Mesenchymal stem cells (MSC), 35, 78, 81
Mouse hepatitis virus, 15
Multimodality imaging, 187

N
N-cadherin, 83
NF-κB, 142–143
Niches, 78, 81
NOD/Shi mice, 58
NOD-SCID, 97, 135, 138–139
NOD-SCID-IL2Rg-/-(NOG), 97, 99, 101
NOG, 133, 139–143
NOG mouse, 180, 191–194
Nude mice, 54

O
Opportunistic pathogens, 16, 19
Optical imaging systems, 184
Organoids, 41
Osteoblastic niche, 78
Ovalubmin (OVA), 98

P
Pasteurella pneumotropica, 16–17
Perforin null mice, 56
Peripheral blood lymphocytes (PBL), 57
PET/CT, 187, 194
PMA/ionomycin, 100
Pneumocystis carinii, 15, 17
Positron emission tomography (PET), 179–180, 183–184, 187–190
Pseudomonas aeruginisa, 16–17, 19

Index

Q
Q-dot, 180, 186–187

R
R5 HIV-1, 135–137, 140
RAG-2/ γ_c^{null}, 139, 141
Rag2–/– γc–/–mice, 125
Rag2^{null} mice, 57
Reaggregate thymic organ culture (RTOC), 101
Regenerative medicine, 39
RNAi, 68

S
S100A4, 173, 174
Severe combined immunodeficiency (SCID) 97
  BLT or NOD/SCID-hu thy/liv, 153–163
  NOD/SCID, 151–153, 158, 161
  SCID, 149, 151–152, 161
  SCID-hu thy/liv, 149, 151–154, 161, 163
SCID-hu thy/liv, 134–135, 140
Self-renewal, 78
Single photon emission computed tomography (SPECT), 179–180, 184, 187–190
Single positive (SP), 99
SPECT/CT, 187, 194
Speed congenic, 12
Staphylococcus aureus, 15, 17
Stem cell culture, 33
Stromal cell-derived factor, 83
Superantigens, 98, 158, 163

T
T-cell
  development, 125
  homeostasis, 150
  human MHC-restriction, 150, 152, 156–158, 162–163
  naive/memory phenotype, 154, 157
  neogenesis/thymopoiesis, 149–150, 152, 156, 161
  reactivity, 127
  thymic organoid, 151, 155
T cell receptor (TCR), 100
T cell repertoire, 155
T-cell selection, 127
TgPVR, 7
Thrombopoiesis, 66
Thrombopoietin, 67
Thymoma, 60
Thymus-repopulating cells, 79
Toxic shock syndrome toxin-1 (TSST-1), 101, 158–160, 162
Transgenic mice, 32

U
Ultrasound imaging 179, 181, 187–190

V
Vascular niche, 78
V(D)J rearrangement, 55

X
X4 HIV-1, 135–137, 140
Xenograft 97
XID mice, 54
X-ray resistance, 60

Current Topics in Microbiology and Immunology
Volumes published since 2002

Vol. 271: **Koehler, Theresa M. (Ed.):** Anthrax. 2002. 14 figs. X, 169 pp. ISBN 3-540-43497-6

Vol. 272: **Doerfler, Walter; Böhm, Petra (Eds.):** Adenoviruses: Model and Vectors in Virus-Host Interactions. Virion and Structure, Viral Replication, Host Cell Interactions. 2003. 63 figs., approx. 280 pp. ISBN 3-540-00154-9

Vol. 273: **Doerfler, Walter; Böhm, Petra (Eds.):** Adenoviruses: Model and Vectors in VirusHost Interactions. Immune System, Oncogenesis, Gene Therapy. 2004. 35 figs., approx. 280 pp. ISBN 3-540-06851-1

Vol. 274: **Workman, Jerry L. (Ed.):** Protein Complexes that Modify Chromatin. 2003. 38 figs., XII, 296 pp. ISBN 3-540-44208-1

Vol. 275: **Fan, Hung (Ed.):** Jaagsiekte Sheep Retrovirus and Lung Cancer. 2003. 63 figs., XII, 252 pp. ISBN 3-540-44096-3

Vol. 276: **Steinkasserer, Alexander (Ed.):** Dendritic Cells and Virus Infection. 2003. 24 figs., X, 296 pp. ISBN 3-540-44290-1

Vol. 277: **Rethwilm, Axel (Ed.):** Foamy Viruses. 2003. 40 figs., X, 214 pp. ISBN 3-540-44388-6

Vol. 278: **Salomon, Daniel R.; Wilson, Carolyn (Eds.):** Xenotransplantation. 2003. 22 figs., IX, 254 pp. ISBN 3-540-00210-3

Vol. 279: **Thomas, George; Sabatini, David; Hall, Michael N. (Eds.):** TOR. 2004. 49 figs., X, 364 pp. ISBN 3-540-00534X

Vol. 280: **Heber-Katz, Ellen (Ed.):** Regeneration: Stem Cells and Beyond. 2004. 42 figs., XII, 194 pp. ISBN 3-540-02238-4

Vol. 281: **Young, John A. T. (Ed.):** Cellular Factors Involved in Early Steps of Retroviral Replication. 2003. 21 figs., IX, 240 pp. ISBN 3-540-00844-6

Vol. 282: **Stenmark, Harald (Ed.):** Phosphoinositides in Subcellular Targeting and Enzyme Activation. 2003. 20 figs., X, 210 pp. ISBN 3-540-00950-7

Vol. 283: **Kawaoka, Yoshihiro (Ed.):** Biology of Negative Strand RNA Viruses: The Power of Reverse Genetics. 2004. 24 figs., IX, 350 pp. ISBN 3-540-40661-1

Vol. 284: **Harris, David (Ed.):** Mad Cow Disease and Related Spongiform Encephalopathies. 2004. 34 figs., IX, 219 pp. ISBN 3-540-20107-6

Vol. 285: **Marsh, Mark (Ed.):** Membrane Trafficking in Viral Replication. 2004. 19 figs., IX, 259 pp. ISBN 3-540-21430-5

Vol. 286: **Madshus, Inger H. (Ed.):** Signalling from Internalized Growth Factor Receptors. 2004. 19 figs., IX, 187 pp. ISBN 3-540-21038-5

Vol. 287: **Enjuanes, Luis (Ed.):** Coronavirus Replication and Reverse Genetics. 2005. 49 figs., XI, 257 pp. ISBN 3-540- 21494-1

Vol. 288: **Mahy, Brain W. J. (Ed.):** Foot-and-Mouth-Disease Virus. 2005. 16 figs., IX, 178 pp. ISBN 3-540-22419X

Vol. 289: **Griffin, Diane E. (Ed.):** Role of Apoptosis in Infection. 2005. 40 figs., IX, 294 pp. ISBN 3-540-23006-8

Vol. 290: **Singh, Harinder; Grosschedl, Rudolf (Eds.):** Molecular Analysis of B Lymphocyte Development and Activation. 2005. 28 figs., XI, 255 pp. ISBN 3-540-23090-4

Vol. 291: **Boquet, Patrice; Lemichez Emmanuel (Eds.):** Bacterial Virulence Factors and Rho GTPases. 2005. 28 figs., IX, 196 pp. ISBN 3-540-23865-4

Vol. 292: **Fu, Zhen F. (Ed.):** The World of Rhabdoviruses. 2005. 27 figs., X, 210 pp. ISBN 3-540-24011-X

Vol. 293: **Kyewski, Bruno; Suri-Payer, Elisabeth (Eds.):** CD4+CD25+ Regulatory T Cells: Origin, Function and Therapeutic Potential. 2005. 22 figs., XII, 332 pp. ISBN 3-540-24444-1

Vol. 294: **Caligaris-Cappio, Federico, Dalla Favera, Ricardo (Eds.):** Chronic Lymphocytic Leukemia. 2005. 25 figs., VIII, 187 pp. ISBN 3-540-25279-7

Vol. 295: **Sullivan, David J.; Krishna Sanjeew (Eds.):** Malaria: Drugs, Disease and Post-genomic Biology. 2005. 40 figs., XI, 446 pp. ISBN 3-540-25363-7

Vol. 296: **Oldstone, Michael B. A. (Ed.):** Molecular Mimicry: Infection Induced Autoimmune Disease. 2005. 28 figs., VIII, 167 pp. ISBN 3-540-25597-4

Vol. 297: **Langhorne, Jean (Ed.):** Immunology and Immunopathogenesis of Malaria. 2005. 8 figs., XII, 236 pp. ISBN 3-540-25718-7

Vol. 298: **Vivier, Eric; Colonna, Marco (Eds.):** Immunobiology of Natural Killer Cell Receptors. 2005. 27 figs., VIII, 286 pp. ISBN 3-540-26083-8

Vol. 299: **Domingo, Esteban (Ed.):** Quasispecies: Concept and Implications. 2006. 44 figs., XII, 401 pp. ISBN 3-540-26395-0

Vol. 300: **Wiertz, Emmanuel J.H.J.; Kikkert, Marjolein (Eds.):** Dislocation and Degradation of Proteins from the Endoplasmic Reticulum. 2006. 19 figs., VIII, 168 pp. ISBN 3-540-28006-5

Vol. 301: **Doerfler, Walter; Böhm, Petra (Eds.):** DNA Methylation: Basic Mechanisms. 2006. 24 figs., VIII, 324 pp. ISBN 3-540-29114-8

Vol. 302: **Robert N. Eisenman (Ed.):** The Myc/Max/Mad Transcription Factor Network. 2006. 28 figs., XII, 278 pp. ISBN 3-540-23968-5

Vol. 303: **Thomas E. Lane (Ed.):** Chemokines and Viral Infection. 2006. 14 figs. XII, 154 pp. ISBN 3-540-29207-1

Vol. 304: **Stanley A. Plotkin (Ed.):** Mass Vaccination: Global Aspects – Progress and Obstacles. 2006. 40 figs. X, 270 pp. ISBN 3-540-29382-5

Vol. 305: **Radbruch, Andreas; Lipsky, Peter E. (Eds.):** Current Concepts in Autoimmunity. 2006. 29 figs. IIX, 276 pp. ISBN 3-540-29713-8

Vol. 306: **William M. Shafer (Ed.):** Antimicrobial Peptides and Human Disease. 2006. 12 figs. XII, 262 pp. ISBN 3-540-29915-7

Vol. 307: **John L. Casey (Ed.):** Hepatitis Delta Virus. 2006. 22 figs. XII, 228 pp. ISBN 3-540-29801-0

Vol. 308: **Honjo, Tasuku; Melchers, Fritz (Eds.):** Gut-Associated Lymphoid Tissues. 2006. 24 figs. XII, 204 pp. ISBN 3-540-30656-0

Vol. 309: **Polly Roy (Ed.):** Reoviruses: Entry, Assembly and Morphogenesis. 2006. 43 figs. XX, 261 pp. ISBN 3-540-30772-9

Vol. 310: **Doerfler, Walter; Böhm, Petra (Eds.):** DNA Methylation: Development, Genetic Disease and Cancer. 2006. 25 figs. X, 284 pp. ISBN 3-540-31180-7

Vol. 311: **Pulendran, Bali; Ahmed, Rafi (Eds.):** From Innate Immunity to Immunological Memory. 2006. 13 figs. X, 177 pp. ISBN 3-540-32635-9

Vol. 312: **Boshoff, Chris; Weiss, Robin A. (Eds.):** Kaposi Sarcoma Herpesvirus: New Perspectives. 2006. 29 figs. XVI, 330 pp. ISBN 3-540-34343-1

Vol. 313: **Pandolfi, Pier P.; Vogt, Peter K. (Eds.):** Acute Promyelocytic Leukemia. 2007. 16 figs. VIII, 273 pp. ISBN 3-540-34592-2

Vol. 314: **Moody, Branch D. (Ed.):** T Cell Activation by CD1 and Lipid Antigens, 2007, 25 figs. VIII, 348 pp. ISBN 978-3-540-69510-3

Vol. 315: **Childs, James, E.; Mackenzie, John S.; Richt, Jürgen A. (Eds.):** Wildlife and Emerging Zoonotic Diseases: The Biology, Circumstances and Consequences of Cross-Species Transmission. 2007. 49 figs. VII, 524 pp. ISBN 978-3-540-70961-9

Vol. 316: **Pitha, Paula M. (Ed.):** Interferon: The 50th Anniversary. 2007. VII, 391 pp. ISBN 978-3-540-71328-9

Vol. 317: **Dessain, Scott K. (Ed.):** Human Antibody Therapeutics for Viral Disease. 2007. XI, 202 pp. ISBN 978-3-540-72144-4

Vol. 318: **Rodriguez, Moses (Ed.):** Advances in Multiple Sclerosis and Experimental Demyelinating Diseases. 2008. XIV, 376. ISBN 978-3-540-73679-9

Vol. 319: **Manser, Tim (Ed.):** Specialization and Complementation of Humoral Immune Responses to Infection. 2008. XII, 174. ISBN 978-3-540-73899-2

Vol. 320: **Paddison, Patrick J.; Vogt, Peter K. (Eds.):** RNA Interference. 2008. VIII, 273. ISBN 978-3-540-75156-4

Vol. 321: **B. Beutler (Ed.):** Immunology, Phenotype First: How Mutations Have Established New Principles and Pathways in Immunology. 2008. ISBN 978-3-540-75202-250

Vol. 322: **Romeo, Tony (Ed.):** Bacterial Biofilms. 2008. XII, 299. ISBN 978-3-540-75417-6

Vol. 323: **S. Tracy; M. S. Oberste; K. M. Drescher (Eds.):** Group B Coxsackieviruses. 2008. ISBN 978-3-540-75545-6